Treatment Wetlands

Biological Wastewater Treatment Series

The *Biological Wastewater Treatment* series is based on the book *Biological Wastewater Treatment in Warm Climate Regions* and on a highly acclaimed set of best-selling textbooks. This international version is comprised of seven textbooks giving a state-of-the-art presentation of the science and technology of biological wastewater treatment.

Titles in the *Biological Wastewater Treatment* series are:

Volume 1: Wastewater Characteristics, Treatment and Disposal
Volume 2: Basic Principles of Wastewater Treatment
Volume 3: Waste Stabilisation Ponds
Volume 4: Anaerobic Reactors
Volume 5: Activated Sludge and Aerobic Biofilm Reactors
Volume 6: Sludge Treatment and Disposal
Volume 7: Treatment Wetlands

Biological Wastewater Treatment Series

VOLUME SEVEN

Treatment Wetlands

Written by:

IWA Task Group on Mainstreaming the Use of Treatment Wetlands

Gabriela Dotro, Günter Langergraber, Pascal Molle, Jaime Nivala, Jaume Puigagut, Otto Stein, Marcos von Sperling

IWA PUBLISHING

DESA UFMG

Published by IWA Publishing, Alliance House, 12 Caxton Street, London SW1H 0QS, UK

Telephone: +44 (0) 20 7654 5500; Fax: +44 (0) 20 7654 5555; Email: publications@iwap.co.uk

Website: **www.iwapublishing.com**

First published 2017

Copy-edited and typeset by Nova Techset, Chennai, India.

British Library Cataloguing in Publication Data
A CIP catalogue record for this book is available from the British Library

Library of Congress Cataloguing in Publication Data
A catalogue record for this book is available from the Library of Congress

ISBN: 9781780408767 (paperback)

ISBN: 9781780408774 (ebook)

Contents

Acronym list

Acronym	Full text
ABR	Anaerobic Baffled Reactors
Al	Aluminium
AS	Activated Sludge
BOD_5	5-day Biochemical Oxygen Demand
Ca	Calcium
COD	Chemical Oxygen Demand
CSO	Combined Sewer Overflow
CSTR	Continuous Flow Stirred-Tank Reactor
DO	Dissolved Oxygen
EPNAC	Evaluation des Procédés Nouveaux d'Assainissement des petites et moyennes Collectivités
ET	Evapotranspiration
Fe	Iron
FWS	Free Water Surface
HF	Horizontal Flow
HLR	Hydraulic Loading Rate
HRT	Hydraulic Residence Time
HSSF	Horizontal Subsurface Flow
LCA	Life Cycle Assessment
MFC	Microbial Fuel Cell
N	Nitrogen
NH_4-N	Ammonium Nitrogen
O&M	Operation and Maintenance
P	Phosphorus

Acronym	Full text
PE	Population Equivalent
PEM	Proton Exchange Membrane
pH	Potential of Hydrogen
PLC	Programmable Logic Controller
PO_4-P	Phosphate Phosphorus
Redox	Oxidation-reduction
SBR	Sequencing Batch Reactor
TIS	Tanks-in-Series
TKN	Total Kjeldahl Nitrogen
TN	Total Nitrogen
TP	Total Phosphorus
TS	Total Solids
TSS	Total Suspended Solids
TW	Treatment Wetland
UASB	Upflow Anaerobic Sludge Blanket
UK	United Kingdom
USA	United States of America
VF	Vertical Flow
VS	Volatile Solids

Foreword

The book "*Biological Wastewater Treatment in Warm Climate Regions*" was written by Marcos von Sperling and Carlos Chernicharo, both from the Federal University of Minas Gerais, Brazil. It was published in 2005 by IWA Publishing, with the main objective of presenting in a balanced way theory and practice of wastewater treatment, so that a conscious selection, design and operation of wastewater treatment processes could be practiced. Theory is considered essential for the understanding and autonomous use of the working principles of wastewater treatment. Practice is associated to the direct application of the material for conception, design and operation. In order to ensure the practical and didactic view of the book, a large number of illustrations, summary tables and design examples were included. Besides being used as a textbook at academic institutions, it was seen that the book was an important reference for practising professionals, such as engineers, biologists, chemists and environmental scientists, acting in consulting companies, water authorities and environmental agencies.

Because the book was very large (two volumes, with a total of around 1,500 pages), it was later on decided to give another alternative to readers, and publish it as a series of books. In 2007 the text was then released by IWA Publishing as six separate books, comprising the "*Biological Wastewater Treatment Series*". The titles that comprise the series are listed in this book cover and preface.

Recognising that the content of the books should reach a wider readership, especially from developing countries, who have more difficulties in purchasing international material, the authors asked IWA Publishing to also make the books available for free downloading, by anyone, anywhere. This open-access format for a book was a pioneering initiative within IWA Publishing, recognising its worldwide reach and the importance of supporting sanitation initiatives in less developed countries. From 2013, both the book "*Biological Wastewater Treatment in Warm Climate Regions*" and the "*Biological Wastewater Treatment Series*" have been available as open-access. The books can be downloaded at http://www.iwapublishing.com/open-access-ebooks/3567.

Throughout this time, the authors felt that the books were missing an important content, related to constructed wetlands for wastewater treatment, a very important process for both developed and developing counties, and warm and temperate climates. It was then very fortunate when the *IWA Task Group on Mainstreaming the Use of Treatment Wetlands* of the *IWA Specialist Group on Wetland Systems for Water Pollution Control* decided to add another volume to the series. With "*Treatment Wetlands*", the series of books now comprises seven volumes. A team of top experts in treatment wetlands prepared this excellent contribution to the series.

This new book keeps the same format, approach and objectives of the previous books. However, in order to keep consistency with the international literature on treatment wetlands and to facilitate the reader in cross-referencing from different sources, there are some differences (for instance, in notation and nomenclature). This book has a more worldwide view, covering not only warm climate regions, but also temperate and cold climates, from where most of the current existing knowledge on research and application of treatment wetlands originates.

I would like to extend a warm appreciation to all those involved in this new project. I am convinced that this new open-access addition to the series will bring an effective contribution to the wastewater sector, and will cater for the dissemination of this important treatment technology on a worldwide basis, with emphasis on countries whose sanitation infrastructure is strongly dependent on simple, effective and affordable wastewater treatment technologies.

Marcos von Sperling

Coordinator of the "Biological Wastewater Treatment Series"

August 2017

Preface

This volume on treatment wetlands is intended to be an addition to the *Biological Wastewater Treatment* Series that is available as a free eBook online at IWA Publishing: http://www.iwapublishing.com/open-access-ebooks/3567. The series now contains seven volumes:

1. Wastewater Characteristics, Treatment and Disposal
2. Basic Principles of Wastewater Treatment
3. Waste Stabilisation Ponds
4. Anaerobic Reactors
5. Activated Sludge and Aerobic Biofilm Reactors
6. Sludge Treatment and Disposal
7. Treatment Wetlands

The target audience of this volume on treatment wetlands is bachelor students with basic knowledge on biological wastewater treatment, as well as practitioners seeking general information on the use of treatment wetlands. This volume focusses on the main types of treatment wetlands for domestic wastewater applications and does not aim to replace any of the current treatment wetland textbooks, including:

- Kadlec R.H. and Wallace S.D. (2009) *Treatment Wetlands, Second Edition*. CRC Press, Boca Raton, FL, USA.

- Vymazal J., and Kröpfelová L. (2008) *Wastewater Treatment in Constructed Wetlands with Horizontal Sub-Surface Flow*. Springer: Dordrecht, The Netherlands.

- Wallace S.D., and Knight R.L. (2006) *Small-scale constructed wetland treatment systems: Feasibility, design criteria, and O&M requirements*. Final Report, Project 01-CTS-5, Water Environment Research Foundation (WERF), Alexandria, Virginia, USA.

- Kadlec, R.H., and Knight R.L. (1996) *Treatment Wetlands*. CRC Press, Boca Raton, FL, USA.

The authors of this volume thank Tom Headley for writing Section 8.4 (Floating treatment wetlands). Jan Vymazal is kindly acknowledged for providing material for Section 3.5 (Horizontal flow wetland case study). Karin Tonderski is kindly acknowledged for providing material for Section 7.4 (Free water surface case study).

The authors also thank the reviewers from the IWA Specialist Group *"Wetland Systems for Water Pollution Control"* for their support and feedback.

List of authors

Authors:

IWA Task Group on Mainstreaming the Use of Treatment Wetlands

Gabriela Dotro, Cranfield University, United Kingdom
Günter Langergraber, BOKU University Vienna, Austria
Pascal Molle, IRSTEA, France
Jaime Nivala, Helmholtz Center for Environmental Research – UFZ, Germany
Jaume Puigagut, UPC Barcelona, Spain
Otto Stein, Montana State University, USA
Marcos von Sperling, Federal University of Minas Gerais, Brazil

List of expert reviewers:

Dirk Esser, SINT, France
Ana Galvão, Instituto Superior Técnico, Portugal
Fabio Masi, IRIDRA S.r.l., Italy
Clodagh Murphy, ARM Ltd, UK
Christoph Platzer, Rotária do Brasil Ltda., Brazil
Bernhard Pucher, BOKU University Vienna, Austria
Anacleto Rizzo, IRIDRA S.r.l., Italy
Diederik Rousseau, Ghent University, Belgium
Alexandros Stefanakis, BAUER Resources GmbH, Germany

Structure of this VOLUME 7 on TREATMENT WETLANDS

Chapter 1 *Overview*, which outlines the main treatment wetland designs and treatment wetland applications considered within the context of this volume.

Chapter 2 *Fundamentals of treatment wetlands*, which summarises wetland specific fundamentals regarding pollutant and pathogen removal processes, hydraulics and energy balance, wetland kinetics, and design considerations.

Chapter 3 *Horizontal flow wetlands*, which introduces HF wetlands by showing their main applications, describing their main processes, discussing their design, listing operation and maintenance requirements, and providing design examples and case studies.

Chapter 4 *Vertical flow wetlands*, which describes VF wetlands in a similar way as HF wetlands in the previous chapter.

Chapter 5 *French vertical flow wetlands*, which presents the French VF wetlands for treating raw wastewater.

Chapter 6 *Intensified wetlands*, which presents the main means of intensification commonly applied for treatment wetlands, including use of reactive media, recirculation, reciprocation, partial saturation, and aeration.

Chapter 7 *Free water surface wetlands*, which describes FWS wetlands, primarily used for tertiary treatment when dealing with domestic wastewater.

Chapter 8 *Other applications*, which describes applications of treatment wetlands besides treating domestic wastewater including zero discharge wetlands, combined sewer overflow wetlands, sludge treatment wetlands, floating treatment wetlands, and microbial fuel cell treatment wetlands.

Chapter 9 *Additional aspects*, which describes important aspects such as process based numerical models, micropollutant removal, economic assessment, and environmental assessment.

References Includes a complete list of references used in the text.

1

Overview of treatment wetlands

Treatment wetlands are natural treatment technologies that efficiently treat many different types of polluted water. Treatment wetlands are engineered systems designed to optimise processes found in natural environments and are therefore considered environmentally friendly and sustainable options for wastewater treatment. Compared to other wastewater treatment technologies, treatment wetlands have low operation and maintenance (O&M) requirements and are robust in that performance is less susceptible to input variations. Treatment wetlands can effectively treat raw, primary, secondary or tertiary treated sewage and many types of agricultural and industrial wastewater. This volume focuses on domestic wastewater treatment using treatment wetlands.

Treatment wetlands can be subdivided into surface flow and subsurface flow systems. Although there are many wetland variants in the literature, in this volume a simple approach is adopted, and four treatment wetlands are primarily discussed (Figure 1.1).

Subsurface flow treatment wetlands are subdivided into Horizontal Flow (HF) and Vertical Flow (VF) wetlands depending on the direction of water flow. In order to prevent clogging of the porous filter material, HF and VF wetlands are generally used for secondary treatment of wastewater. VF wetlands for treating screened raw wastewater have also been introduced and successfully applied. These so-called French VF wetlands provide integrated sludge and wastewater treatment in a single system and thus save on construction costs, because primary treatment of wastewater is not required. Free Water Surface (FWS) wetlands (also known as surface flow wetlands) are densely vegetated units, in which the water flows above the media bed. In subsurface flow wetlands, the water level is kept below the surface of a porous medium such as sand or gravel. FWS wetlands are generally used for tertiary wastewater treatment.

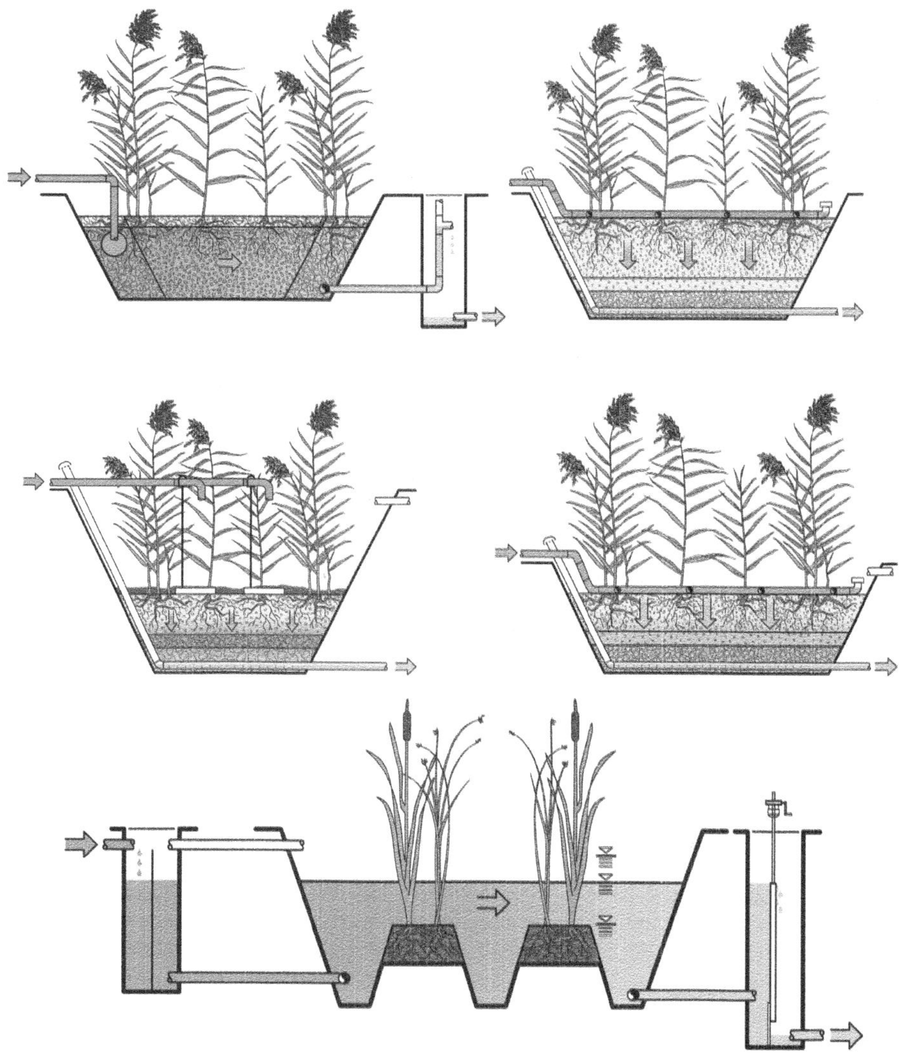

Figure 1.1 Overview schematics of treatment wetlands addressed in this volume. Top left: horizontal flow; top right: vertical flow; middle left: French vertical flow, first stage; middle right: French vertical flow, second stage; bottom: free water surface.

Table 1.1 presents a summary of the four main treatment wetland types covered in this volume: HF wetlands, VF wetlands, French VF wetlands and FWS wetlands.

Table 1.1 Main treatment wetland types covered in this volume.

Type	Short description
HF wetland	• Wastewater flows horizontally through a sand or gravel-based filter whereby the water level is kept below the surface. • Due to the water-saturated condition mainly anaerobic degradation processes occur. • Effective primary treatment is required to remove particulate matter to prevent clogging of the filter. • Emergent plants (macrophytes) are used. • Are used for secondary or tertiary treatment.
VF wetland	• Wastewater is intermittently loaded on the surface of the filter and percolates vertically through the filter. • Between two loadings air re-enters the pores and aerates the filter so that mainly aerobic degradation processes occur. • Effective primary treatment is required to remove particulate matter to prevent clogging of the filter. • Emergent macrophytes are used.
French VF wetland	• Are VF wetlands for treating screened wastewater. • Two stages of wetlands operate in series and in parallel. • Provide integrated sludge and wastewater treatment in a single step. • No primary treatment unit is required.
FWS wetland	• Resemble natural wetlands in appearance. • Require large surface area, are generally lightly loaded. • Various plant genus can be used: a) emergent: *Typha, Phragmites, Scirpus*, (b) submerged: *Potamogeon, Elodea*, etc, (c) floating: *Eichornia* (water hyacinth), *Lemna* (duckweed). • Are mainly used for tertiary treatment.

Table 1.2 summarises removal efficiencies that can be expected for typical designs of the four main treatment wetland types. For each of the four main types a great number of modifications exist that can result in higher removal efficiencies.

Table 1.3 compares specific treatment area requirements per population equivalent (PE) of selected technologies for secondary treatment of domestic wastewater. It should be noted that technologies listed in Table 1.3 do not result in the same effluent quality. Anaerobic ponds and upflow anaerobic sludge blanket (UASB) reactors are not frequently used in temperate climates for domestic wastewater treatment but have wider application in warmer climates.

Table 1.2 Typical removal efficiencies of main treatment wetland types.

Parameters	**HF**	**VF** [a]	**French VF**	**FWS**
Treatment step (main application)	Secondary	Secondary	Combined primary and secondary	Tertiary
Total Suspended Solids	> 80%	> 90%	> 90%	> 80%
Organic matter (measured as oxygen demand)	> 80%	> 90%	> 90%	> 80%
Ammonia nitrogen	20 – 30%	> 90%	> 90%	> 80%
Total nitrogen	30 – 50%	< 20%	< 20%	30 – 50%
Total phosphorus (long term)	10 – 20%	10 – 20%	10 – 20%	10 – 20%
Coliforms	2 $\log_{10}$	2 – 4 $\log_{10}$	1 – 3 $\log_{10}$	1 $\log_{10}$

[a] Single-stage VF bed, main layer of sand (grain size 0.06 – 4 mm)

Table 1.3 Land requirement of selected wastewater treatment technologies for secondary treatment for warm to temperate climates.

Treatment technology	**Treatment area requirement (m^2/PE)**
Facultative pond [a]	2.0 – 6.0
Anaerobic + facultative pond [a]	1.2 – 3.0
UASB reactor [a]	0.03 – 0.10
Activated sludge, SBR [a]	0.12 – 0.30
Trickling filter [a]	0.15 – 0.40
HF wetlands [b]	3.0 – 10.0
VF wetlands [b]	1.2 – 5.0
French VF wetlands [c]	2.0 – 2.5

[a] (von Sperling, 2007a)

[b] for warm (Hoffmann *et al.*, 2011) and temperate climates (Kadlec and Wallace, 2009)

[c] for temperate climates (Molle *et al.*, 2005)

Compared to other treatment systems, treatment wetlands have a larger land requirement, but less requirement of external energy and O&M. If the landscape allows, treatment wetlands can be operated without pumps and thus without any external energy input. Like all extensive systems, treatment wetlands are robust and tolerant against fluctuating influent flow and concentration. Treatment wetlands are thus particularly suitable for use as small decentralised treatment systems.

2

Fundamentals of treatment wetlands

2.1 POLLUTANT AND PATHOGEN REMOVAL PROCESSES

Treatment wetlands are complex wastewater treatment systems possessing a diverse set of pollutant and pathogen removal pathways. Unlike other conventional wastewater treatment systems in which removal processes are optimised by a series of separate unit operations designed for a specific purpose, multiple removal pathways simultaneously take place in one or two reactors. Wetland plants play several important roles in treatment wetlands. Primarily, their roots and rhizomes provide attachment sites for microbial biofilms increasing the biological activity per unit area compared to open water systems such as ponds. They diffuse the flow, limiting hydraulic short-circuiting, and can also release small amounts of oxygen and organic carbon compounds into the rooting matrix, fuelling both aerobic and anoxic microbial processes. Indeed, a unique feature of TWs is their ability to support a diverse consortium of microbes; obligate aerobic, facultative, and obligate anaerobic microorganisms can be found due to large redox gradients, a factor contributing to the robust performance of a TW. The heterogeneous distribution of redox conditions within a TW is caused by several factors, especially the presence of the macrophyte root system and, in VF and certain other systems, fluctuations in water level caused by cyclical flow regimes. Major removal pathways within TWs are listed for specific constituents in Table 2.1.

Table 2.1 Main mechanisms for pollutant and pathogen removal in treatment wetlands.

Parameter	Main removal mechanisms
Suspended solids	Sedimentation, filtration
Organic matter	Sedimentation and filtration for the removal of particulate organic matter, biological degradation (aerobic and/or anaerobic) for the removal of dissolved organic matter
Nitrogen	Ammonification and subsequent nitrification and denitrification, plant uptake and export through biomass harvesting
Phosphorus	Adsorption-precipitation reactions driven by filter media properties, plant uptake and export through biomass harvesting
Pathogens	Sedimentation, filtration, natural die-off, predation (carried out by protozoa and metazoa)

Organic matter

Organic matter can be classified and measured in many ways as described in previous chapters in both Volume 1 and Volume 2 (von Sperling, 2007a; von Sperling, 2007b) of this series. Particulate organic matter and soluble organic matter are both considered as inputs. Removal mechanisms for particulate and soluble organic matter differ and depend on treatment wetland design. Generally, Chemical Oxygen Demand (COD) is used as the main analytical method for measuring organic matter, however, 5-day (carbonaceous) Biochemical Oxygen Demand (BOD_5) can also be used. The basic removal mechanisms for dissolved organic matter are the microbial pathways as described in Chapters 2 and 3 of Volume 2 (von Sperling, 2007b), but unlike most wastewater treatment systems, several pathways can be utilised within different micro-sites of the same wetland reactor.

Particulate organic matter

Particulate organic matter entering with the influent wastewater is retained mainly by physical processes such as filtration and sedimentation. The retained particulates accumulate and undergo hydrolysis, generating an additional load of dissolved organic compounds that can be further hydrolysed or degraded within the treatment bed. Particulate organic matter accumulation in the granular medium is a typical feature of subsurface flow treatment wetlands and is one of the main factors behind the operational problem of clogging in these systems. Other sources of particulate organic matter include biofilm growth and plant and microbial detritus accumulation. The relative contribution of the various fractions of particulate organic matter accumulation depends on the applied wastewater load and the properties of the plants and biofilms growing in the system. Overall, particulate organic matter accumulation in subsurface flow treatment wetlands is much higher than the typical particulate influent loading rate, indicating that other materials (such as dead plant material) contribute to particulate organic matter retained within the treatment bed.

Soluble organic matter

Specific microbial pathways for soluble organic matter removal are discussed in Chapter 3 of Volume 2 (von Sperling, 2007b). To review, microbes induce a chemical reaction in which electrons are transferred from organic matter (the electron donor) to a specific compound (the electron acceptor), in the process releasing energy for cell growth. The specific pathway is usually defined by the electron acceptor. The major pathways active in treatment wetlands, listed in decreasing energy release include: aerobic respiration, with oxygen as the electron acceptor and carbon dioxide as the major product; denitrification with nitrate and nitrite as the electron acceptor and nitrogen gas and carbon dioxide as the major products; sulphate reduction with sulphate as the electron acceptor and sulphide and carbon dioxide as the major products; and methanogenesis, in which organic matter is simultaneously the electron acceptor and donor, and carbon dioxide and methane are the primary products. Each pathway has an optimal redox potential and therefore may be active in different locations within the same wetland as there are strong redox gradients as a function of level of saturation and distance from the water surface and plant roots, ranging from strongly anaerobic (less than –100 mV) to fully aerobic (greater than +400 mV).

Aerobic microbial respiration

Many heterotrophic bacteria use oxygen as a terminal electron acceptor, and because it is the pathway with the highest energy yield these microbes will dominate when oxygen is available. Most heterotrophic bacteria are facultative, meaning that they can also use nitrate or nitrite as electron acceptors when oxygen is limiting. Oxygen availability varies greatly among different wetland configurations. Most VF wetlands are operated with pulsed, intermittent surface loading, aerating the bed between pulses and increasing the presence of oxygen in the bulk water. Thus, aerobic respiration is the dominant removal pathway in VF systems. HF wetlands are almost always fully saturated to within a few centimetres of the surface. In HF wetlands, there are only a few sources of oxygen (a) inputs by the influent; (b) physical surface re-aeration, and (c) plant release. Oxygen demand for typical domestic wastewater is much higher than the sum of all these inputs, thus whilst some heterotrophic respiration undoubtedly takes place especially near roots of plants, other pathways are usually dominant. Surface reaeration is greater in FWS than in HF wetlands due to the open water surface, thus whilst more heterotrophic activity is possible less energetically favourable processes are likely dominating, especially in and near the sediments at the bottom.

Denitrification

Denitrification is the biologically mediated reduction of nitrate to nitrogen gas through several intermediary steps in the absence of dissolved molecular oxygen. Under these anoxic conditions and when nitrate is available, denitrification can be a predominant organic matter degradation pathway in TWs, especially in HF wetlands (García *et al.*, 2004). Denitrification has been shown to account for a large fraction of total organic carbon removal in HF wetlands (Baptista *et al.*, 2003). However, nitrate availability is often problematic as it is typically not present in appreciable quantities in the influent and cannot be generated by autotrophic nitrification until sufficient organic matter has been removed.

Sulphate reduction

Sulphate is a common constituent of many types of wastewater and can be used as a terminal electron acceptor by a large group of anaerobic heterotrophic microorganisms. The main product is sulphide, which is a source of nuisance odours and can cause inhibition of both microbial activities and plant growth (Wießner *et al.*, 2005). On the other hand, most metal sulphides are highly insoluble and sulphate reduction is an important metals removal mechanism in TWs. Sulphate reduction can be a very significant organic matter removal pathway, accounting for a substantial fraction of total COD removal in a HF treatment wetland (García *et al.*, 2004).

Methanogenesis

Methanogenesis is an anaerobic microbial reaction in which organic matter is oxidised to carbon dioxide and reduced to methane. Whilst not a strict removal mechanism in terms of COD, the low solubility of methane in water effectively removes the organic matter by outgassing of methane. Required redox conditions for methanogenesis are very similar to those required for sulphate reduction. Furthermore, methanogens and sulphate reducers use similar organic substrates (such as acetic acid and methanol). When the COD-to-sulphate ratio (expressed as COD:S) is lower than 1.5, sulphate-reducing bacteria typically outcompete methanogens and when the ratio is greater than six, methanogens typically predominate (Stein *et al.*, 2007a). At intermediate ratios, the two processes often occur simultaneously and unless the products are of concern, the net effect on organic matter removal can be combined.

Nitrogen

Nitrogen exists in many forms and various interrelated processes convert it from one form to another in a complex system called the nitrogen cycle. Nitrogen enters most primary and secondary treatment wetlands as organic N and ammonium (NH_4-N), with tertiary systems receiving a mixture of nitrogen species including nitrate. In most wetlands, some level of nitrogen transformation is expected and/or mandated before discharge of the final effluent to a water body. In many cases the expectation is conversion to nitrate, a less toxic form of nitrogen, but increasingly more jurisdictions expect total nitrogen (TN) removal from wastewater. Virtually all pathways of the nitrogen cycle are active in treatment wetlands, including mineralisation (ammonification), ammonia volatilisation, nitrification, denitrification, plant and microbial uptake, nitrogen fixation, nitrate reduction, anaerobic ammonia oxidation, adsorption, desorption, burial, and leaching (Vymazal, 2007). However, it is believed that only some of these pathways contribute significantly to the nitrogen transformations and removal mechanisms important in wastewater treatment. It is widely accepted that microbially-induced transformations of nitrogen common to other wastewater treatment systems dominate in treatment wetlands, with sorption and plant uptake also present to a limited extent. The contribution of each pathway is affected by the treatment wetland type, applied loading rate, hydraulic residence time (HRT), temperature, vegetation type and the properties of the medium (Kuschk *et al.*, 2003; Akratos and Tsihrintzis, 2007). The most critical processes are highlighted in this section.

Ammonification

Ammonification consists of the conversion of organic N to ammonium through extracellular activity from enzymes excreted by microorganisms (Vymazal, 2007). Ammonification is considered a necessary first step to nitrogen conversion to nitrate and/or removal, but is seldom a limiting step for subsequent TN removal.

Nitrification

Nitrification is the oxidation of ammonium to nitrate facilitated by a consortium of autotrophic microbes with nitrite as a major intermediate product. For the process to take place, the microorganisms, oxygen, alkalinity and micronutrients must be present in the wastewater. Autotrophic nitrifiers are typically slower growing microorganisms than aerobic heterotrophs and can thus be outcompeted in the presence of readily biodegradable organic matter. A major advantage of VF wetlands is their high oxygenation capacity and thus their ability to nitrify. Some nitrification can occur in HF systems, especially when lightly loaded with organic matter, but nitrification is often a limiting step to nitrogen removal in HF systems. Nitrification alone is a conversion process and does not result in nitrogen removal, unless it is adequately coupled to denitrification.

Denitrification

Denitrification was discussed as an organic carbon removal mechanism but is vital to effective nitrogen removal as it converts nitrate to nitrogen gas that is released to the atmosphere. Denitrification is often difficult to achieve in secondary treatment wetlands (and most wastewater treatment systems in general) because the nitrification process is typically a prerequisite to convert the influent ammonia into nitrate, which cannot take place until the sufficient organic carbon is consumed. This can result in insufficient residual organic matter for denitrification. The high oxygenation potential of VF systems makes them poor at denitrification, as the process requires anoxia to end in the generation of nitrogen gas. Most HF systems will denitrify all of the nitrate that can be created within if used for secondary treatment, whereas partial denitrification is more common in tertiary systems. Therefore, VF wetlands typically remove little TN but have high concentrations of nitrate in the effluent. HF wetlands can remove TN to some extent, but the effluent can still contain high concentrations of ammonium nitrogen. Environmental factors known to influence denitrification rates include the level of dissolved oxygen (DO), pH, redox potential, type of media and the organic matter concentration, among others.

Sorption

Ammonium is a cation and is therefore readily sorbed onto media particles within treatment wetlands. Sorption may be near 100% of the influent for a short time after start-up of a wetland system. However, the sorption capacity of all media is finite and once all sites are saturated very little additional sorption can take place (Vymazal, 2007). The capacity is controlled by the media particle size and chemical composition. The capacity is usually much smaller for the sand and gravel particles typical in treatment wetlands than for natural soils because the available surface area per unit volume of wetland is comparatively small. Specific media with a higher ammonium sorption capacity, such as zeolite, can be used to prolong the sorption capacity of the system.

Though sorption is a minor removal mechanism, it can aid the nitrification-denitrification removal process in TW systems that are loaded intermittently by temporarily storing ammonium, allowing time for heterotrophs to consume most of the organic matter, then exposing the sorbed ammonium to oxygen during the waiting period. Nitrification can then take place. Upon the next dose, the nitrate can react with the new dose of organic matter, allowing denitrification to take place and restoring the sorption site for a new molecule of ammonium. Extremely high sorptive capacities or very low loading rates are required for this mechanism to dominate operation of intermittent systems such as VF wetlands.

Plant uptake

A common misconception is that plants remove most of the nitrogen in treatment wetlands. Emergent macrophytes do store nitrogen in their tissue and plant uptake results in nitrogen removal ranging from 0.2 to 0.8 g $N/m^2 \cdot d$, depending on the macrophyte species considered (Vymazal, 2007). Some of this stored nitrogen can be removed by regular harvesting of above ground biomass, however more than half of the nitrogen uptake may be stored in below ground tissue and timing is important as plants translocate nitrogen between above and below ground tissue depending on the season. Harvesting is also an operational cost and its cost effectiveness is questionable unless the system is lightly loaded. If plants are not harvested, no net nitrogen removal occurs because any nitrogen in plant tissue is eventually released during decomposition of the plant matter.

Phosphorus

Phosphorus enters most treatment wetlands primarily as organic phosphorus and orthophosphate, but most organic phosphorus is converted to orthophosphate as part of organic matter degradation. Mechanisms that play a part in phosphorus removal in treatment wetlands include chemical precipitation, sedimentation, sorption and plant and microbial uptake. Unfortunately, most of these processes are slow or not active unless special media are used to enhance abiotic processes. As with nitrogen, plants incorporate phosphorus into their biomass but this can be a removal mechanism only if plants are harvested and is thus subject to the same limitations as nitrogen plant uptake as a removal mechanism.

The effectiveness of treatment wetlands for phosphorous removal is determined by the applied loading rate. In very lightly loaded FWS systems, such as for effluent polishing, phosphorus removal can be excellent and due primarily to soil accretion (sedimentation and co-precipitation with other minerals). For treatment of typical secondary wastewater using VF and HF systems, removal is generally quite modest once the sorptive capacity of the media is saturated. Considerable research has been conducted to find media with high phosphorus sorptive capacities with some success. These filter media are referred to as reactive media (see Section 6.2). As all media, reactive media have a finite capacity, however, it is possible to delay saturation to a period of years, which may be suitable in certain situations. Another option is to use an additional unplanted filter bed in which the reactive media can be periodically replaced without losing the removal capacity for other constituents in upstream cells. This sacrificial filter is generally left unplanted to facilitate removal of the material once it reaches its sorption capacity. A common approach is to dose chemical salts (iron or aluminium based) to react with phosphorus upstream of the treatment wetland and use the system to retain any residual precipitated solids (Brix and Arias, 2005; Lauschmann *et al.*, 2013; Dotro *et al.*, 2015).

Pathogens

Pathogen removal in TWs is extremely complex due to the variety of processes that may lead to the removal or inactivation of bacteria, viruses, protozoans or parasites. Treatment wetland technology offers a suitable combination of physical, chemical and biological mechanisms required to remove pathogenic organisms. The physical factors include filtration and sedimentation, and the chemical factors include oxidation and adsorption to organic matter. The biological removal mechanisms include oxygen release and bacterial activity in the root zone (rhizosphere), as well as aggregation and retention in biofilms, natural die-off, predation, and competition for limiting nutrients or trace elements.

Most of the available data concerning the capacity of treatment wetlands to remove pathogens is focused on faecal indicator organisms; less information is available for specific bacteria, viruses, protozoan oocysts and other parasites such as helminth eggs. Removal of indicator organisms in treatment wetlands is dependent on the type of wetland system, the operational conditions and the characteristics of influent wastewater. It is generally accepted that conventional subsurface treatment wetland designs can remove up to 3 $\log_{10}$ units of faecal bacteria indicators, but the relative importance of specific removal mechanisms is still unknown.

2.2 WATER AND ENERGY BALANCES

Water and energy balance considerations must be made for any wastewater treatment unit. Whilst the fundamental concepts of mass balance and reactor hydraulics introduced in Chapter 2 of Volume 2 of this book series (von Sperling, 2007b) are unchanged, this section is intended to highlight how features unique to a treatment wetland affect these budgets. The large surface area of most treatment wetlands requires consideration of the water fluxes in addition to the inflow and outflow of the system. Treatment wetlands share this feature with wastewater ponds but macrophytes alter the magnitude of some fluxes. It is often convenient to separate these fluxes from the wastewater flow by borrowing terminology from the field of hydrology. The large scale of a treatment wetland also makes it prudent to assess water flow within the wetland separately from the reactor hydraulics discussed in Section 2.3. Finally, the energy flux between the wetland and its surrounding environment is important to some hydrologic fluxes and to prevent freezing during winter operation in cold regions. This chapter discusses hydrology, hydraulics, and energy balances specific to treatment wetlands.

Hydrologic budget

Treatment wetlands are often used to treat stormwater or the excess flow of domestic wastewater during rain events in combined sewer overflows (CSOs). The importance of a hydrologic budget to proper design is obvious in these cases as the majority of the water in the wetland is responding to rain falling remotely from the wetland. But hydrology must be considered even in cases where the treatment wetland is designed to treat exclusively domestic wastewater. There are several water fluxes that must be considered in addition to the influent and effluent, as shown in Figure 2.1.

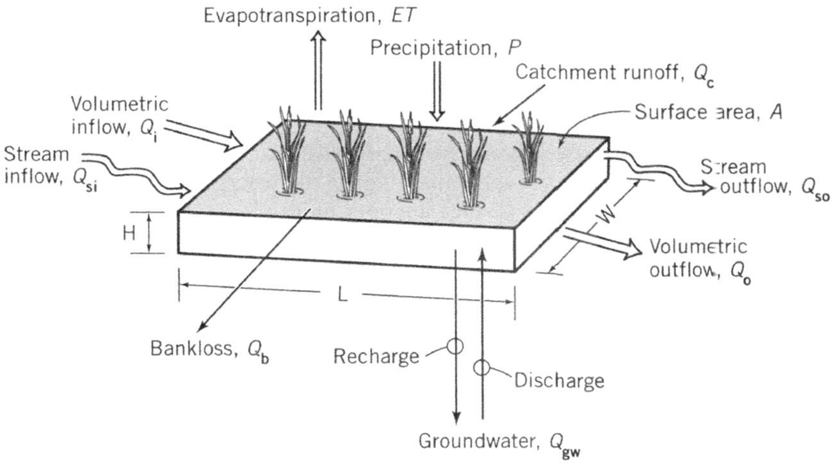

Figure 2.1 Water fluxes in a treatment wetland. Reprinted with permission from Kadlec and Wallace (2009).

A hydrologic budget is represented by Equation 2.1 (Kadlec and Wallace, 2009).

$$Q_i - Q_o + Q_c - Q_b - Q_{gw} + Q_{sm} + (P \times A) - (ET \times A) = \frac{dV}{dt} \tag{2.1}$$

where:

A = wetland surface area, m^2

ET = evapotranspiration rate, m^3/d

P = precipitation rate, m/d

Q_b = bank loss rate, m^3/d

Q_c = catchment runoff rate, m^3/d

Q_{gw} = infiltration rate to groundwater, m^3/d

Q_i = wastewater influent flow rate, m^3/d

Q_o = wastewater effluent flow rate, m^3/d

Q_{sm} = snowmelt rate, m^3/d

t = time, d

V = water volume in wetland, m^3

It should be noted that notation here is slightly different from that adopted in the previous books of this series. However, it was decided to keep them as such, to make them compatible with classical wetland literature, and facilitate the reader in cross-referencing from different sources. For instance, influent and effluent here are denoted by subscripts "i" and "o", meaning "input" and "output", or "inlet" and "outlet" – e.g. Q_i and Q_o. In the other volumes of this book series, these are treated as Q_0 and Q_e ("0" standing for zero, or entrance, and "e" for effluent).

Consistent units must be used, with particular consideration to flow rates that are usually presented as volume per unit time (e.g. m^3/d) whereas the other fluxes are usually presented as volume per surface area per unit time (or length per unit time, e.g. $m^3/m^2{\cdot}d$ or m/d). The major design goal is to ensure the wetland maintains a water level between a minimum and maximum. As most treatment wetlands are lined, infiltration to soil and embankments is usually negligible, but infiltration may be an important flux in unlined wetlands. In such scenarios, the groundwater infiltration rate Q_w can also be (seasonally) positive, where the water table rises and enters the wetland. Precipitation records are available for almost any location. During the design process, it is key to pick appropriate storm durations and return intervals to determine the appropriate precipitation design depth. Generally, return intervals greater than ten years are sufficient. An appropriate storm duration should consider the responsiveness of the outlet to changes in water depth, but could be conservatively set at the theoretical HRT of the wetland. All wetlands should be designed with some freeboard (e.g., water depth above the maximum design depth) to temporarily store precipitation events larger than the design storm.

In TWs for domestic wastewater the water balance can be simplified, by assuming steady-state conditions (no water accumulation *dV/dt*) and by not considering the terms related to stream flow (in and out) and catchment run-off. Furthermore, for design purposes, if the wetland unit is lined (no infiltration to groundwater), effluent flow can be simply estimated by: influent flow + direct precipitation over wetland surface – evapotranspiration (*ET*) (Equation 2.2 and Figure 2.2):

$$Q_o = Q_i + (P \times A) - (ET \times A) \tag{2.2}$$

where:

Q_0 = wastewater effluent flow rate, m^3/d

Q_i = wastewater inflow rate, m^3/d

P = precipitation rate, m/d

A = wetland surface area, m^2

ET = Evapotranspiration rate, m^3/d

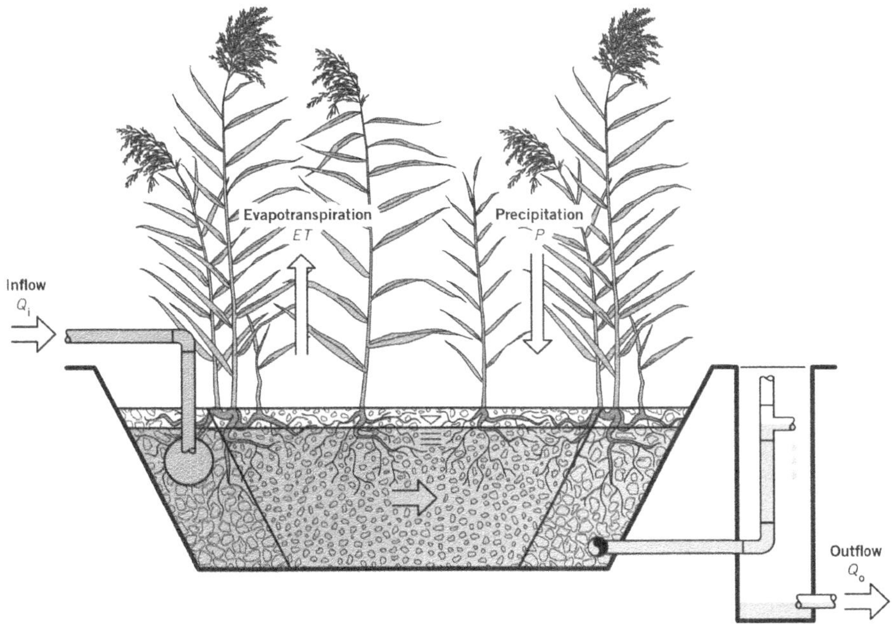

Figure 2.2 Simplified water balance in a treatment wetland receiving wastewater and subject to only evapotranspiration and precipitation.

Evapotranspiration from a wetland is not well quantified. Large FWS wetlands are likely to behave similarly to an open water body of similar size, thus the ET rate is typically estimated as a fraction of the pan evaporation rate. The pan evaporation coefficient varies slightly with regional climatic factors but a value of 0.8 is typical (McCuen, 2016). Estimating ET for small wetlands and especially those with subsurface flow is more difficult. Plant transpiration dominates the ET flux and small surface areas increase the "edge effect" as more plant canopy is exposed to the solar, wind and other energy inputs driving ET. Rates of ET in these systems can exceed pan evaporation rates and factors such as aspect ratio and orientation relative to the general wind direction can have significant impact on the ET rate. The ET rate can also vary significantly throughout the year depending on the developmental stage of the plant and the plant species selected. In fact, some treatment wetlands have been designed as zero-discharge systems even in locations where rainfall exceeds ET on an annual basis by maximising ET losses using large aspect ratios perpendicular to the prevailing wind direction (see Section 8.1). A simple pan evaporation analysis to estimate the magnitude of ET relative to the other fluxes can be made as a first step, and if the ET component is found to be significant relative to precipitation and influent, a more detailed energy balance approach is necessary.

Once the fluxes into a wetland are estimated, water flow through the TW is determined by application of two fundamental concepts: conservation of mass (continuity) and conservation of momentum. Application of the momentum equation is different for FWS, HF and unsaturated VF systems. A power law similar to Manning's Equation is usually applied to FWS wetlands, and Darcy's Law is usually applied to HF systems. The Richards equation is usually applied to unsaturated VF systems and is beyond the scope of this text, but readers are directed to Freeze and Cherry (1979) for more information. The use of these equations is discussed in the next subsection.

Continuity

The superficial horizontal velocity of flow *u* through a wetland is related to the flow rate *Q* and cross-sectional area of flow via the continuity equation (Equation 2.3).

$$u = \frac{Q}{wh} \tag{2.3}$$

where:

Q = flow rate, m^3/d

w = width of the wetland perpendicular to flow, m

h = wetland water depth, m

u = horizontal velocity at the longitudinal axis, m/d

For subsurface flow wetlands, the actual velocity of the liquid through the pores may be calculated incorporating the medium porosity in the denominator of Equation 2.3.

The width and especially depth may not be constant in large wetlands and flow rate is typically not constant because the effects of precipitation and ET are distributed down the wetland flow length. Also, incoming wastewater flow is variable during the day. However, the equation is valid at a specific cross section and can be averaged over the wetland by using the average of the influent and effluent flow rate. In other words, for some calculations, the flow can be considered to be:

$$Q = \frac{Q_i + Q_o}{2} \tag{2.4}$$

However, taking into account that the designer stipulates the influent flow but, in many cases, has difficulties in doing a proper water balance and thus in estimating the effluent flow, in most situations, the flow used in design calculations is the influent flow Q_i. This is the approach adopted in this volume, unless otherwise stated. For existing treatment wetlands, where influent and effluent flows can be easily measured, the average of both flows can be used, as shown in Equation 2.4.

FWS wetland momentum

Because flow in a FWS wetland is almost always laminar, a typical semi-empirical momentum equation for turbulent flow such as Manning's Equation is not valid because Manning's *n* is not a constant in these conditions. In addition, frictional losses include the effects of plants and litter, as well as bed and sidewall friction. Thus a simple power function has been proposed to account for friction in the momentum equation in FWS wetlands (Kadlec and Wallace, 2009) (Equation 2.5).

$$u = a\,h^{(b-1)} S^{c} \qquad (2.5)$$

where:

a,b,c = friction parameters, unitless

u = superficial water velocity, m/d

h = wetland water depth, m

$S = -dH/dx$, = negative of the water surface slope, m/m

Values of b and c are often assumed to be three and one, respectively, with a increasing from 1.0×10^7 to 5×10^7 m/d as vegetation density decreases. Equations 2.1 and 2.5 can be solved simultaneously to determine the appropriate slope for an average depth and flow under steady, semi-uniform conditions. Kadlec and Wallace (2009) provide a more detailed hydraulic analysis when variation in depth along the length of the FWS wetland is important.

HF wetland momentum

Flow through a horizontal subsurface flow wetland is best described by Darcy's Law (Equation 2.6).

$$u = k_e \frac{dh}{dx} \qquad (2.6)$$

where:

u = superficial water velocity, m/d

h = wetland water depth, m

k_e = effective hydraulic conductivity, m/d

dh/dx = water surface slope, m/m

Equations 2.1 and 2.6 can be simultaneously solved for a bed slope that is equal to the hydraulic grade line, thus ensuring a constant depth along the length, provided the hydraulic conductivity is appropriately determined. A schematic of the hydraulic profile in a horizontal subsurface flow constructed wetland with flat bottom (no slope) is shown in Figure 2.3.

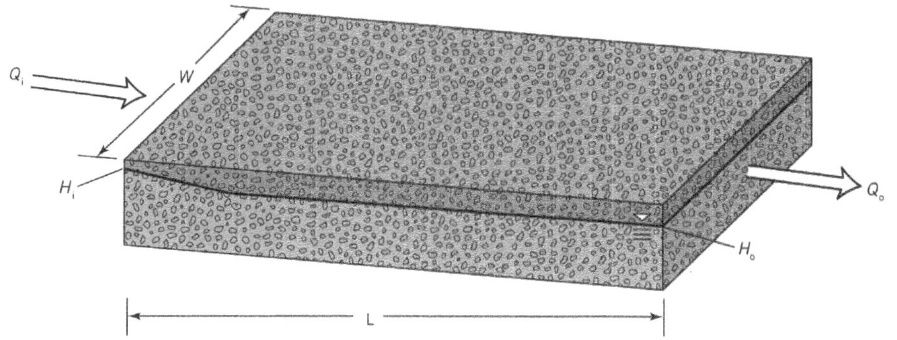

Figure 2.3 Schematic longitudinal section of a horizontal subsurface flow constructed wetland. Simplified image reprinted with permission from Kadlec and Wallace (2009).

Unfortunately, it is difficult to measure hydraulic conductivity in treatment wetlands (Knowles and Davies, 2009; Knowles *et al.*, 2010; Matos *et al.*, 2017). The hydraulic conductivity of the bed medium is highest at start-up, when the medium is relatively clean, and decreases with time as plant roots, microbial biofilms and chemical precipitates gradually occupy spaces that were initially filled with water. Many early HF wetland designs considered only the initial hydraulic conductivity, which as pores accumulated solids forced the flow to the surface and often resulted in serious short-circuiting and a sharp decrease in treatment performance. Conversely, designing for lower long-term hydraulic conductivity allows water to flow too rapidly through the porous medium, which decreases the depth of water toward the tail of the wetland under initial conditions. Plant establishment in this zone is then hindered. Current HF wetland designs use the lower long-term estimate of k_e to ensure water will always remain beneath the surface of the gravel. In some systems, a water depth regulator is installed in the outlet works in order to maintain a shallower slope when hydraulic conductivity is greater than the estimated clean-bed design value.

Figure 2.4 shows schematic views of the increase in head loss that occurs with the progressive clogging in the system, eventually leading to surface overland flow when the head loss h_f becomes higher than the medium freeboard (distance between the top of the bed and the water level at the beginning of operation). When this happens, Darcy's Law equation for flow in porous media is no longer valid, because of the presence of overland flow. Clogging is considered the major operational issue in subsurface flow wetlands. Considerable research effort is dedicated to the investigation of the clogging development, control and remediation (Knowles *et al.*, 2011; Nivala *et al.*, 2012).

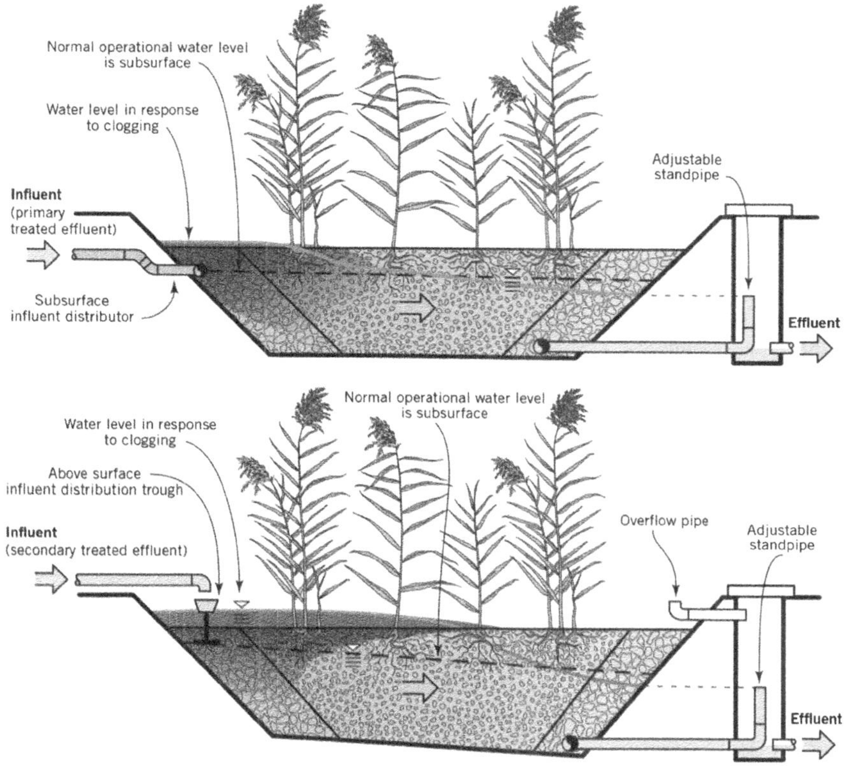

Figure 2.4 Responses to clogging on HF wetlands; top: secondary HF system; bottom: tertiary HF system.

Hydraulic residence time and hydraulic loading rates

The HRT, also referred to as hydraulic retention time or hydraulic detention time, is an important parameter when considering hydraulics of a reactor in combination with the kinetics of chemical and microbial transformations in treatment wetlands (see Section 2.3). The HRT is defined as the time a molecule of water stays in the wetland on average, from entrance to exit, and is typically calculated as the water volume in the reactor divided by the flow rate. However, flow through a wetland varies in space and time and the volume may or may not consider the volume of the wetland filled with plant material and media and may or may not consider "dead spaces" due to hydraulic inefficiency. In HF wetlands, the water volume of the wetland may be only 30 to 45% (taking into account the porosity of the media) of the nominal volume of the wetland basin itself (length × width × depth) (Kadlec and Wallace, 2009). Often the flow rate is assumed to be the average of the inflow and outflow; however, in many cases only inflow is measured and is therefore often used to determine the HRT.

Selection of an appropriate volume is more problematic. Using only the pore volume occupied by water is more appropriate than using the volume of the wetland basin itself. However, it must be recognised that porosity varies with time, and is seldom measured and is almost never known exactly. It is thus important to define the variables used when calculating HRT of a treatment wetland.

Another factor influencing the HRT and treatment performance of a TW is the hydraulic efficiency. Flow is never uniform across the width and depth of a wetland so that water may remain longer in certain locations and move faster through others. A good design includes features to minimise this short-circuiting and ensure good mixing within the wetland (Wahl *et al.*, 2010). The analysis of residence time distribution functions originating from tracer studies is one of the main tools for the assessment of hydraulic performance of wetlands. In order to simplify the analysis, hydraulic indexes extracted from these functions are normally used to characterise short-circuiting and mixing behaviour (Teixeira and Siqueira, 2008).

In this text, the theoretical HRT (or τ) in saturated treatment wetlands is defined using influent flow (Q_i) and the estimated water volume of the wetland (taking into account the porosity of the media in subsurface flow wetlands, as illustrated in Figure 2.5) unless explicitly stated otherwise (Equation 2.7). One should note that the void space is also occupied by a biofilm layer with biomass growth around the medium and by the roots of the plants. The term "theoretical" HRT is used, because it cannot be guaranteed that the actual HRT will be the same as the calculated. As a matter of fact, imperfections in the hydraulic behaviour inside TWs, with the presence of dead zones, short circuits and other factors lead to the fact that the actual retention time of each individual water molecule is smaller than the theoretical one, given by Equation 2.7.

$$\tau = \frac{\text{Liquid volume}}{\text{Flow}} = \frac{\varepsilon V}{Q_i} = \frac{\varepsilon h A}{Q_i} \qquad (2.7)$$

where:

τ = nominal (theoretical) hydraulic retention time, d

ε = porosity (fraction of wetland volume occupied by water), unitless

h = wetland water depth, m

A = wetland surface area, m^2

Q_i = influent flow rate, m^3/d

In a FWS wetland, since there is no support medium, porosity is taken as 1.0 in Equation 2.7, and the theoretical HRT can be calculated by the usual expression of V/Q_i.

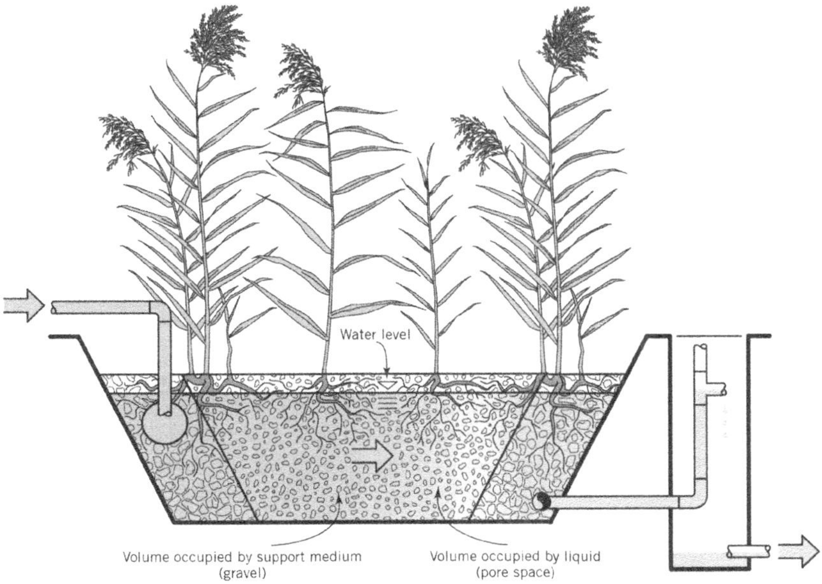

Figure 2.5 Schematic representation of a subsurface flow constructed wetland showing the volumes occupied by liquid and by the support medium.

Treatment wetland surface hydraulic loading rate (HLR_s or q) in this volume is based on influent flow Q_i, and thus q is defined as shown in Equation 2.8:

$$HLR_S = q = \frac{Q_i}{A} \tag{2.8}$$

where:

$HLR_S = q =$ surface hydraulic loading rate, $m^3/(m^2 \cdot d)$

$A =$ wetland surface area, m^2

$Q_i =$ influent flow rate, m^3/d

The concept of volumetric hydraulic loading rate (HLR_v) is also used in wastewater treatment. It has the units of $m^3/m^3 \cdot d$, or m^3/d of influent wastewater divided by m^3 of reactor volume, as shown in Equation 2.9.

$$HLR_V = \frac{Q_i}{V} = \frac{Q_i}{h \times A} \tag{2.9}$$

where:

HLR_V = volumetric hydraulic loading rate, $m^3/(m^3 \cdot d)$

Q_i = influent flow rate, m^3/d

V = reactor volume, m^3

A = wetland surface area, m^2

h = wetland water depth, m

It should be noted that A in Equation 2.8 and V in Equation 2.9 are the actual wetland surface area and volume, and do not incorporate porosity. It should be mentioned that the concept of HRT, as calculated by Equation 2.7, is only valid for saturated media, such as the one in HF wetlands. With VF wetlands with intermittent pulse dosing, the void spaces in the bed are occupied by air in the interval between feeding batches. Therefore, there is no concept of liquid volume in the bed and HRT cannot be calculated for this type of vertical wetland. However, the concept of surface HLR (Equation 2.8), and volumetric HLR (Equation 2.9) can be used for both VF wetlands and HF wetlands.

Energy budget

An energy budget is important in order to assess ET rates when a preliminary estimate determines that it is an important flux in the wetland hydrologic budget. Another reason to apply an energy budget is to prevent or manage freezing of a wetland in winter in cold climates. FWS wetlands can be operated year-round in temperate climates provided the water level is increased in the autumn to maintain an ice-free layer underneath. Freezing of a HF or VF wetland will prevent water from flowing through the porous medium. If this occurs, there is no option but to wait until the system thaws, which is problematic.

The major energy fluxes of a treatment wetland are shown in Figure 2.6. If the wetland is warmer than the air, heat is lost from the perspective of the wetland but is gained when the air temperature is warmer. Similarly, energy is received from the ground if the wetland is colder than the surrounding soil, and heat is delivered to the ground if the wetland is warmer. In summer, the net solar radiation is sufficient to warm the wetland so that the two reversible fluxes G and E_{loss} exit the wetland. In temperate climates, net solar radiation is greatly reduced during winter and the air temperature is often below freezing. In this case, the major energy inputs are from the influent wastewater and ground-heat transfer. In either season, Wallace and Knight (2006) define a "balance point" temperature near the inlet, which remains through the rest of the wetland. During subfreezing air temperatures, the balance point temperature in the wetland can approach zero. It is important to use the energy balance method in order to ensure the balance point temperature remains above freezing in winter, and (if applicable), does not exceed any temperature-related discharge standards in summer.

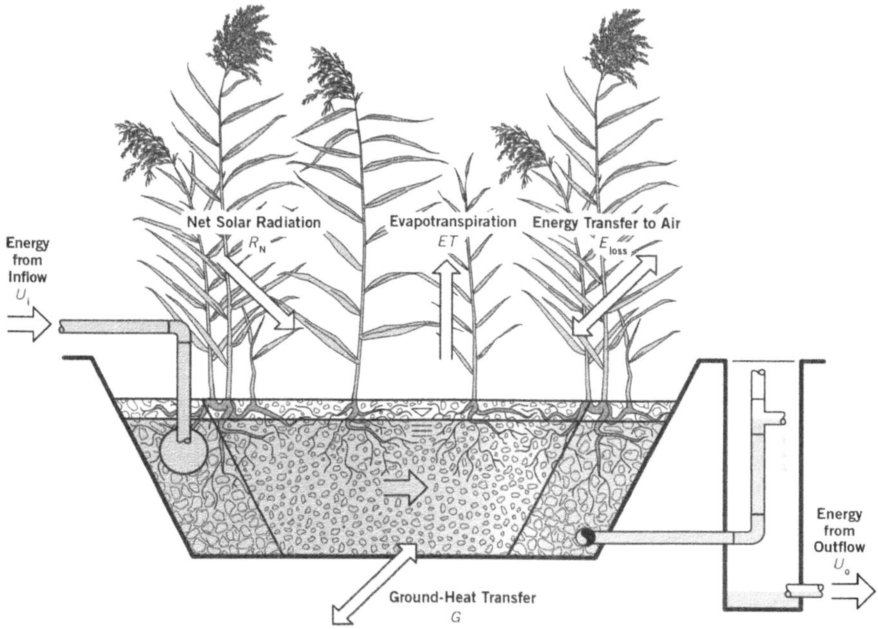

Figure 2.6 Energy fluxes in a treatment wetland. Modified from Wallace and Knight (2006).

In a FWS wetland, ice formation provides a significant insulation barrier to heat loss to the atmosphere and the wetland can be managed for an expected ice thickness by raising the water level to maintain the appropriate HRT for proper treatment in the flow under the ice. The colder the ambient air temperature, the thicker the ice must be to maintain free-flowing water underneath. The water level in FWS wetlands must be kept relatively constant during winter operation in order to prevent possible uprooting of plant stems and leaves encased in the ice layer, thus the appropriate ice thickness must be determined in advance. In an established HF or VF wetland, the layer of thatch and plant detritus on the surface provides an insulating blanket (which can be further augmented by snow cover), especially if the wetland is designed with additional freeboard to collect blowing snow. In the first year or two after start-up of a HF or VF wetland, the plants may not be well enough established and freezing may be encountered. To avoid this, a layer of mulch can be added to provide the necessary insulation blanket and water level can be dropped a few centimetres to provide an extra layer of air-filled void space in the gravel to minimise the heat loss to the air. Wallace and Knight (2006) provide methods for determining appropriate ice thickness in FWS wetlands and mulch types and thicknesses to mitigate freezing in subsurface flow wetlands.

2.3 KINETICS AND REACTOR HYDRAULICS

Chemical and biological degradation in wastewater treatment is often represented by chemical engineering reaction kinetics. For a detailed description of reaction kinetics and mass balances, refer to Volume 2, Sections 2.2 and 2.3 of this book series (von Sperling, 2007b). In addition to understanding the fundamentals of reaction kinetics and mass balances, a basic understanding of reactor hydraulics is also necessary. VF and French VF wetlands generally behave differently and are sized using other methods. The reader is referred to Chapter 4 (VF wetlands) and Chapter 5 (French VF wetlands) for further detail.

Reactor hydraulics

The fundamentals of reactor hydraulics are discussed in Volume 2, Section 2.4 of this book series (von Sperling, 2007b). A summary of the most frequently used hydraulic models are provided in Volume 2, Table 2.1 (von Sperling, 2007b). The reader is advised to consult this reference because previous knowledge of these hydraulic models is essential to understanding the concepts and equations presented in the current chapter. The fundamentals of reactor hydraulics discussed herein apply only to saturated treatment wetlands (FWS, HF, and saturated aerated wetland designs). Hydraulics of unsaturated treatment wetland designs (VF, French VF) must be described via other methods. Early FWS and HF treatment wetland design guidance used the assumption of ideal plug-flow reactor dynamics (e.g., number of tanks-in-series (TIS) = ∞) (Kadlec and Knight, 1996). However, this plug-flow assumption has proven to be too simplistic and is no longer recommended for describing treatment wetland hydraulics (Kadlec and Wallace, 2009). Treatment wetland hydraulics is best represented by the TIS model, which is an intermediate case between the ideal plug-flow and continuous flow stirred-tank reactor (CSTR) extremes (Wallace and Knight, 2006; Kadlec and Wallace, 2009), see also Volume 2, Section 2.4.4 of this book series (von Sperling, 2007b). The representation of one single reactor by a series of complete-mix tanks is purely for mathematical convenience when using equations for design purposes or to represent an existing system. In either case, the objective is to determine the effluent concentration of this single reactor.

If the reactor is represented by one CSTR (N=1), this means that it is being represented by a perfect, or idealised, complete-mix tank. If, on the other hand, the reactor is represented by an infinite number of CSTR tanks-in-series (N=∞), this means that, theoretically, it behaves as a perfect plug-flow reactor. Of course, these are two extreme boundaries and all wetland units, in practice, will behave in between these two idealised conditions.

Plug-flow and single CSTR equations have been widely used in wastewater treatment. This means that these two idealised hydraulic models are being used to represent non-ideal reactors, and this distorts the values of the kinetic coefficients when they are obtained from measurements of inlet and outlet concentrations. These coefficients cease to be purely kinetic coefficients, and incorporate in themselves the imperfections of the hydraulic behaviour of an actual tank, modelled as an idealised one. This is why it is important to have a good representation of the hydraulic behaviour of the treatment unit, so that a good prediction of the effluent concentration can be made, using reaction coefficients that approach, as close as possible, the true intrinsic kinetic coefficient.

The representation of one wetland unit by a series of CSTR units is an attempt to overcome these difficulties. What one needs to know, besides the value of the reaction coefficient, is the number of *N* tanks to be used in the representation of the wetland unit under study. In existing wetland units, this can be done by tracer tests (methodology not shown here). For further detail on treatment wetland tracer testing, the reader is referred to Appendix B in Kadlec and Wallace (2009).

For design purposes, one needs to adopt the value of the number of TIS (*N*, or also *N*TIS) from the literature, based on similar wetland units. The equivalent number of TIS that best represents one reactor is a function of several factors, with emphasis on the ratio between length (*L*) and width (*W*; the *L*:*W* ratio). The more elongated the wetland, the higher the *L*:*W* ratio, and therefore it is expected that the number of equivalent TIS (*N*TIS) will be higher. Conversely, for a reactor with low *L*:*W* ratio (equal to or lower than one), it is expected that it will be more well mixed, and thus the equivalent *N*TIS will be low.

Table 2.2 summarises the reported hydraulic behaviour as determined by tracer tests of various saturated treatment wetland designs. Since the number of TIS (*N*TIS) is only a mathematical representation of the hydraulic performance of a wetland, it does not need to be an integer value. These values are presented here for the reader to have an idea of common *N*TIS values, but it should always be remembered that the likely number will depend very much on the geometric relationships of the treatment wetland.

Table 2.2 Hydraulic behaviour of treatment wetland designs as determined by tracer testing.

Design	**Tanks-in-Series (*N*TIS)**	**Source**
Horizontal Flow [a]	8.3	Kadlec and Wallace (2009)
Aerated Horizontal Flow	4.5	Boog (2013)
Aerated Vertical Flow	1.1	Boog et al. (2014)
Free Water Surface [b]	3.6	Kadlec and Wallace (2009)

[a] median value from 35 studies
[b] median value from 37 studies

In Volume 2, Section 2.4.5 of this book series (von Sperling, 2007b), the dispersed-flow model (also called plug-flow with dispersion) was also presented. This model is another convenient possibility to represent a real reactor, whose behaviour lies in-between the idealised models of CSTR and plug-flow. The relevant equation for predicting effluent concentration is presented in the cited reference. It incorporates the dispersion coefficient d (dimensionless) which represents the degree of longitudinal dispersion in the reactor. A value of $d = 0$ indicates no longitudinal dispersion, that is, a behaviour of a perfect plug-flow reactor. A value of $d = \infty$ indicates perfect mixing, that is, a hydraulic behaviour of an ideal complete-mix reactor. In practice, all treatment wetlands fall in-between these two idealised regimens. The dispersed-flow model should only be used for units whose dispersion number d is less than 1.0, which accommodates most of the wetland units found in practice. The considerations here are similar to those made for *N*TIS. The dispersion coefficient d may also be determined by tracer tests. Naturally there is a relationship between d and *N*TIS. The lower d, the higher *N*TIS, with the opposite also holding true. There are equations for converting d into *N*TIS, and vice-versa, but they are beyond the scope of this volume. In this volume, preference is given to the TIS model, because it is more widely applied in the representation of treatment wetlands. For other treatment processes, such as stabilisation ponds, the dispersed-flow model has been extensively applied.

The hydraulics of unsaturated treatment wetland designs (VF, French VF, as well as some intensified designs such as reciprocating systems) cannot be adequately described by the TIS model. Any designation of an *N*TIS to unsaturated VF wetland designs is only mathematical in nature. Tracer testing demonstrates that the hydraulics in such wetland designs are not well represented by classical chemical engineering reactor dynamics.

Areal and volumetric rate coefficients

The fundamentals of first-order reaction equations are discussed in detail in Volume 2, Sections 2.2.3 and 2.4.4 of this book series (von Sperling, 2007b). The fundamentals of hydraulics of treatment wetlands with saturated water flow (HF and FWS wetlands) are given in Section 2.2 of this volume. In the past, the traditional equation for plug-flow was frequently used for the design and representation of wetland units (Equation 2.10). This equation is still used in many designs, but one should always remember that actual wetlands do not behave as an idealised plug-flow reactor, and thus the prediction of outlet concentrations may suffer large deviations from reality because of the inadequacy of the hydraulic model to represent a real unit.

$$C_o = C_i e^{-k\tau} \tag{2.10}$$

where:

C_o = outlet concentration, mg/L

C_i = inlet concentration, mg/L

k = first-order reaction coefficient, 1/d

τ = nominal (theoretical) hydraulic retention time, d (Eq. 2.9)

Because of the limitations of the plug-flow model, treatment wetland performance is currently most often described using a modified first-order reaction equation based on non-ideal reactor hydraulics, with the presumption that the TIS are equal in size (Equation 2.11).

$$C_o = \frac{C_i}{(1+k\tau/N)^N} \tag{2.11}$$

where:

C_o = outlet concentration, mg/L

C_i = inlet concentration, mg/L

k = first-order reaction coefficient, 1/d

τ = nominal (theoretical) hydraulic retention time, d

N = number of equivalent tanks in series, dimensionless

N, the number of TIS, is also referred to as *N*TIS in this volume.

In this example, the inlet concentration is C_i=100 mg/L, the first-order reaction coefficient is k = 0.4 d^{-1} and the HRT is τ = 5 d. If one applies Equation 2.11 with these input data, the following values of outlet concentrations C_o will be obtained, for different values of N: N=1, C_o=33mg/L; N=2, C_o=25mg/L; N=5, C_o=19 mg/L; N=10, C_o=16mg/L. The concentration profile along the longitudinal axis is shown in the figure, indicating the subsequent decay as the constituents moves along the wetland (X axis is time, in d). The last compartment indicates the outlet concentration, which is the information required for design purposes.

Equation 2.11 may also be presented in other convenient ways that have the reaction coefficient k as a function of the surface area or wetland volume (Equation 2.12).

$$C_o = \frac{C_i}{(1+k_A/Nq)^N} = \frac{C_i}{(1+k_V\tau/N)^N} \tag{2.12}$$

where:

C_o = outlet concentration, mg/L

C_i = inlet concentration, mg/L

k_A = first-order areal rate coefficient, m/d

k_V = first-order volumetric rate coefficient, 1/d

q = hydraulic loading rate, m/d

N = number of tanks-in-series, dimensionless

τ = nominal hydraulic retention time, d

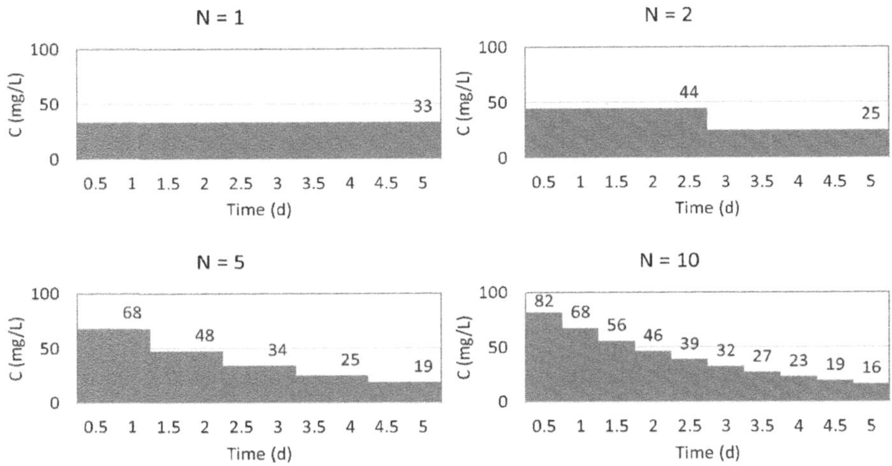

Figure 2.7 Example of the longitudinal profile of concentrations for one horizontal wetland unit that is represented by *N*TIS. Utilisation of Equation 2.11. Input data: C_i=100 mg/L, k = 0.4 d^{-1}, τ = 5 d. Outlet concentration is shown in the last compartment.

One should be careful when interpreting the coefficients k_A and k_V in Equation 2.12. The term k_A/q does not take into account the medium porosity, since q is simply Q_i divided by the total surface area A of the wetland ($q = Q_i/A$). On the other hand, the theoretical HRT τ takes into account the medium porosity ($\tau = V \times \varepsilon/Q_i$). Therefore, when converting one coefficient to the other one should understand that they have different backgrounds, and that medium porosity ε needs to be incorporated in the conversion. The overall coefficient k (Equation 2.11) is similar to k_V (Equation 2.12), and is equivalent to:

$$k = k_V = \frac{k_A}{\varepsilon V/A} = \frac{k_A}{\varepsilon h} \tag{2.13}$$

where:

k = first-order rate coefficient, 1/d

k_V = first-order volumetric rate coefficient, 1/d

k_A = first-order areal rate coefficient, m/d

ε = porosity (fraction of wetland volume occupied by water), unitless

V = wetland volume, m^3

A = wetland surface area, m^2

h = wetland water depth, m

The values of the reaction coefficients k, k_A and k_V do not represent the actual intrinsic kinetic coefficients, as could be determined by batch tests under controlled conditions. They are coefficients based on field data (measurements of inlet and outlet concentrations in existing systems) and, as such, represent both the kinetics and the departures between the hydraulic model assumed and the hydraulic behaviour taking place in reality. They are still useful but one cannot use one reaction coefficient based on one hydraulic model in a different hydraulic model equation. The *N*TIS model aims to give a better representation of the actual wetland hydraulics, and so the associated coefficients are likely to be closer to the intrinsic kinetic coefficients. However, reaction coefficients for the idealised plug-flow model (or for a single CSTR), when obtained from field data, are likely to be very different from the intrinsic kinetic coefficients because the reactor will not be, in practice, equal to the ideal models. All of this has been a matter of substantial confusion in the technical literature when reporting values of kinetic coefficients. Therefore, it is very important that, when stating the value of a reaction coefficient, one specifies the hydraulic model associated with it.

Many ecosystem processes that contribute to pollutant removal in treatment wetlands are dependent on wetland area. Because of this, and due to the fact that early FWS and HF treatment wetland designs did not vary greatly in depth, removal rate coefficients for treatment wetlands have generally been reported on an areal basis (designated by k_A), but this not universal. Table 2.3 shows k_A-rates for HF and FWS wetlands. The values shown in the table are generated from a database of actual treatment wetland performance, showing the percentage of systems that exhibit k_A-rates below the 50% (i.e. 50% systems degraded the pollutant slower than the given value). One should note that k_A values presented in Table 2.3 are expressed at the unit of m/yr, whereas in Equation 2.11 k_A has been presented in m/d. It is just a matter of working with consistent units.

The values shown in Table 2.3 are for wetlands treating primary effluents. For BOD_5 removal under different working conditions, Kadlec and Wallace (2009) report the following k_A values (50th percentile) for HF wetlands:

- Primary effluent: $C_i = 100$ to 200 mg/L; $k_A = 25$ m/yr
- Secondary effluent: $C_i = 30$ to 100 mg/L; $k_A = 37$ m/yr
- Tertiary effluent: $C_i = 3$ to 30 mg/L; $k_A = 86$ m/yr

Table 2.3 Example areal-based reaction rate coefficients (50th percentile) for HF and FWS wetlands (Kadlec and Wallace, 2009).

Pollutant	**HF** k_A-**rate** (m/yr)	**FWS** k_A-**rate** (m/yr)
BOD_5	25	33
TN	8.4	12.6
NH_4-N	11.4	14.7
NO_x-N	41.8	26.5
Thermotolerant coliform	103	83

With the advent of new treatment wetland designs and intensifications, the depth of the treatment wetland can vary greatly from one system to another. When the depth of a wetland varies between systems, it is then necessary to consider removal rate coefficients on a volumetric basis (designated by k_V). Mathematically, an areal rate coefficient can be converted to a volumetric coefficient by dividing by the wetland depth and medium porosity, however caution must be taken if the transformed value is used for a wetland of different depth than the data from which it was created.

Temperature correction factor

Water temperature influences the reaction rates of most pollutant degradation in treatment wetlands. Temperature effects can be described using the Arrhenius temperature equation (Equation 2.14). Equation 2.14 can be used for correcting both areal (k_A) and volumetric (k_V) rate coefficients.

$$k_T = k_{20}\theta^{(T-20)} \tag{2.14}$$

where:

k_T = rate coefficient at water temperature T

k_{20} = rate coefficient at water temperature 20°C

T = water temperature, °C

θ = modified Arrhenius temperature factor, dimensionless

A temperature correction factor of $\theta = 1.0$ indicates that pollutant removal is not influenced by water temperature. A θ-value greater than 1.0 indicates that k increases with increasing water temperature. A θ-value less than 1.0 indicates that *k* decreases with increasing water temperature. Table 2.4 provides average reported temperature correction factors for FWS and HF wetlands (Kadlec and Wallace, 2009). Note that the values in Table 2.4 are from a large set of wetland data. In this table, the values of θ for BOD_5 removal are lower than 1.00, which are different from other biological treatment processes and suggest removal efficiencies could, in fact, deteriorate with an increase in water temperature. This is counterintuitive and in contradiction with reports in the literature, as indeed recognised by Kadlec and Wallace (2009; p258). As such, the recommendation in this volume is to design *without* adjusting the BOD_5 removal rate for temperature. Rate coefficients that have been temperature-corrected are generally referred to as *modified* first-order rate coefficients.

Table 2.4 Example temperature correction factors (θ-values) for HF and FWS wetlands (50th percentile values, Kadlec and Wallace, 2009).

Parameter	HF	FWS
BOD_5	0.981	0.985
TN	1.005	1.056
NH_4-N	1.014	1.014
NO_x-N	–	1.102
Thermotolerant coliform	1.002	–

The influence of water temperature on a rate coefficient can be dramatic. A temperature correction factor (θ) of 1.056, for example, indicates a reduction of 5.6% for each decrease of 1°C. This will lead to a three-fold decrease in rate coefficient (*k*) as the water temperature approaches 0°C. From a design perspective, a three-fold decrease in the rate coefficient would result in a three-fold increase in required wetland area or volume.

Given the strong focus of this book series on warm climate regions, usually the opposite situation occurs: water temperature can be higher than 20°C, and thus reactions will proceed at a faster rate. It is therefore important to have information on the temperature of the water (not air temperature) to be used in the design. In many cases, mean monthly water temperatures of the coldest month are used, to be on the safe side in the design. In other cases, minimum annual water temperatures are used. In this volume, unless otherwise stated, all reaction coefficients are expressed at the standard water temperature of 20°C.

Background concentration

Background concentration (C^*) is an irreducible effluent concentration that results from internal biogeochemical cycling within wetlands. For example, for organic matter, C^* could represent the refractory or non-biodegradable fraction. The background concentration C^*, which is often inferred from a large collection of data, effectively sets a lower limit to the effluent concentration of a treatment wetland (C_o). This means that even for a wetland that has an infinitely long retention time, the theoretical effluent concentration C_o will never be less than C^*. It is especially important to take into account background concentrations when wetland influent concentrations are low ($C_i \leq 3C^*$) or when effluent concentrations approach (or are expected to approach) laboratory detection limits (Kadlec and Wallace, 2009).

Estimates for background concentrations are provided in Table 2.5 (Kadlec and Wallace, 2009). Except for a few parameters (such as BOD_5, COD, and N), C^* for wetlands providing secondary treatment of domestic wastewater (e.g., most HF and VF wetlands) will be close to zero or below laboratory detection limits.

Table 2.5 Example background concentrations (C^*) in mg/L for HF, VF, and FWS wetlands (Kadlec and Wallace, 2009).

Parameter	HF	VF	FWS	
			Lightly Loaded	**Heavily Loaded**
BOD_5	10	2	2	10
TN	1	0	1.5	
NH_4-N	0	0	0.1	0.1

The values shown in Table 2.5 are for wetlands treating primary effluents. For BOD_5 removal, for different influents, Kadlec and Wallace (2009) report the following C^* values (50th percentile) for HF wetlands:

- Primary effluent: C_i = 100 to 200 mg/L; C^* = 10 mg/L
- Secondary effluent: C_i = 30 to 100 mg/L; C^* = 5 mg/L
- Tertiary effluent: C_i = 3 to 30 mg/L; C^* = 1 mg/L

It should be noted that C^* concentrations can also vary with temperature (Stein *et al.*, 2007b). Temperature corrections for C^* can be made by replacing k_T and k_{20} with C^*_T and C^*_{20} in the modified Arrhenius equation (Equation 2.12).

Pollutant weathering

Some wastewater parameters such as COD and BOD_5 provide bulk measurements of a range of organic compounds of varying degradability. Some compounds are more easily (or more quickly) degraded, and others are more difficult (or slower) to degrade. Therefore, the organic matter in the influent wastewater has a different composition than the organic matter that remains in the effluent (Wallace and Knight, 2006). The easiest to degrade organic matter is removed first, meaning that the degradation rate of organic matter decreases with increasing distance and time along the flow path. The decrease in the removal rate can be mathematically represented by modifying (reducing) the number of tanks-in-series (*N*TIS). This modified number of tanks-in-series, denoted as *P*, is the *apparent* number of tanks-in-series. *P* is a fitted (or estimated) parameter and cannot be empirically measured, with the constraint that $P \leq N$ (Kadlec and Wallace, 2009). Note that this hydraulic model is suited only for saturated systems. Any extrapolation to an unsaturated system is purely mathematical in nature. Table 2.6 shows examples of *P* values, but it should be remembered that these are associated with the biodegradability of the constituent and the geometric relationships in the wetland (that define the boundary value of *N*). Therefore, other values may be found in practice.

Table 2.6 Examples of *P* values for HF, VF, and FWS wetlands (Kadlec and Wallace, 2009).

Parameter	HF	VF	FWS
BOD_5	3	2	1
TN	6	n.g. [a]	3
NH_4-N	6	6	3

[a] n.g. = not given

*P-k-C** approach

The most recent kinetic equation for representing pollutant degradation in treatment wetlands is a modified first-order equation with a non-zero background concentration. Treatment wetland performance has been shown to be well represented by the *P-k-C** approach (Equation 2.15, see also Kadlec and Wallace, 2009). Note that Equation 2.15 has the same structure as the traditional equation for the TIS model (Equation 2.11). It simply deducts the fraction of background concentration *C** from the inlet and outlet concentrations, and substitutes *N* by *P*.

$$(C_o - C^*) = \frac{(C_i - C^*)}{(1 + k\tau/P)^P} \qquad (2.15)$$

where:

C_o = outlet concentration, mg/L

C_i = inlet concentration, mg/L

C^* = inlet concentration, mg/L

k = first-order reaction coefficient, 1/d

τ = nominal (theoretical) hydraulic retention time, d

P = apparent number of tanks-in-series (TIS), dimensionless

The outlet concentration C_o can be simply obtained by rearrangement of Equation 2.15:

$$C_o = C^* + \frac{C_i - C^*}{(1 + k\tau/P)^P} \tag{2.16}$$

Equations 2.15 and 2.16 can be also presented with a further detailing of the reaction coefficient, which can be expressed on an areal or volumetric basis (Equation 2.17). Information required for calculating a rate coefficient using the *P-k-C** approach includes the physical attributes of the system (length, width, and effective depth of the treatment cell, as well as the porosity of the porous medium), operational data (flow rate(s), effluent water temperature, influent and effluent pollutant concentrations), as well as estimated parameters (for systems providing secondary treatment of domestic wastewater, P and C^* are often estimated) (Kadlec and Wallace, 2009). See comments made on Equation 2.11 regarding the conversion of k_A into k_V, and vice-versa.

Within the scope of this volume, HLR (q) and HRT (τ) are based on the inflow rate Q_i. This is a simplifying assumption. In reality, factors such as rainfall and ET can greatly affect the overall water balance and HRT of a treatment wetland.

$$\left(\frac{C_o - C^*}{C_i - C^*}\right) = \frac{1}{(1 + k_A/Pq)^P} = \frac{1}{(1 + k_V\tau/P)^P} \tag{2.17}$$

where:

C_o = outlet concentration, mg/L

C_i = inlet concentration, mg/L

C^* = background concentration, mg/L

k_A = modified first-order areal rate coefficient, m/d

k_V = modified first-order volumetric rate coefficient, 1/d

P = apparent number of tanks-in-series (TIS), dimensionless

q = hydraulic loading rate, m/d

τ = hydraulic retention time, d

The *P-k-C** approach is described in great detail in Kadlec and Wallace (2009). There, extensive information is available for HF and FWS wetlands, as well as other design considerations such as risk tolerance, seasonal trends in treatment performance, synoptic error, and stochastic variability. VF and French VF wetlands are sized using other methods. The reader is referred to Chapter 4 (VF wetlands) and Chapter 5 (French VF wetlands) for further detail.

2.4 DESIGN APPROACHES

There are many ways to size and design a treatment wetland system. Over recent decades, TW design approaches have evolved from simple rule-of-thumb to regression-based approaches to more advanced calculations that take into account factors such as HLR, non-ideal flow, background concentration, and pollutant weathering. For any wetland design, it is essential to keep in mind that published design parameters are based on operational data from real-world systems. Kadlec and Wallace (2009) recommend performing a short check to ensure that data extrapolation during the design process is avoided. Equations and/or design parameters should only be applied to new designs that fall within the range of the datasets from which they were derived. The new design should fall within the physical and operational conditions of the source data, including:

- Type of treatment wetland
- Inlet and outlet concentrations
- Hydraulic and mass loadings
- Size, aspect ratio and depth
- Climate and associated water gains and/or losses (rainfall, ET, etc.)
- Ecology and plant community
- Open water fraction (for FWS wetlands only)

The most common design approaches include:

- Rule-of-thumb
- Regression equations
- Plug-flow *k-C**
- Loading charts
- *P-k-C**

From these design approaches, only the rule-of-thumb and loading charts are applicable to VF and French VF wetlands, all other approaches are only applicable to HF and FWS wetlands. The design approaches for VF and French VF wetlands are described in Chapter 4 (VF wetlands) and Chapter 5 (French VF wetlands). With the exception of the "rule-of-thumb" approach, all others consider a specific pollutant (e.g., BOD_5) to be removed for a particular water quality target. In practice, most treatment wetlands are designed to remove multiple pollutants. Like with other treatment technologies, the designer needs to conduct the calculations for all pollutants of interest and select the resulting design that will enable *all* the target pollutants to be removed.

Rule-of-thumb

Rule-of-thumb is a prescriptive design approach based on a particular wetland application in a specific climatic or geographical region. Most often, this approach is used for a single wetland technology (most commonly HF or VF) in a local or national guideline (Brix and Johansen, 2004; Macrophytes et Traitement des Eaux, 2005; DWA, 2017; ÖNORM, 2009). Generally, design advice is given in terms of area requirement per person equivalent (m^2/PE), but can also be given, for example, as an areal loading rate (g $BOD_5/m^2 \cdot d$ or g $COD/m^2 \cdot d$). Table 2.7 presents a selection of rule-of-thumb design recommendations given in area of wetland required per person (m^2/PE). This approach is a practical way of starting a design procedure and can be effective when there is substantial accumulated knowledge on the application of the technology in the region under consideration. However, great care must be taken that these design recommendations are not extrapolated to situations where the boundary conditions (pre-treatment technology, per capita wastewater generation, climate, etc.) differ greatly from those under which the recommendations were created. The references listed in Table 2.7 demonstrate some different rule-of-thumb design recommendations for various wetland types in temperate climates, but is by no means an all-inclusive list.

Table 2.7 Rule-of-thumb design recommendations for temperate climates.

Country	Technology	Specific surface area (m^2/PE)	Reference
Austria	VF	4	ÖNORM B 2505 (2009)
Denmark	HF	5	Brix and Johansen (2004)
	VF	3	
Germany	VF	4	DWA-A 262 (2017)
France	French VF	2	Iwema et al. (2005)

The design values presented here are related to temperate climate countries. For warm climate regions, which are the focus of this series of books, loading rates may be higher and areal requirements are lower. Therefore, it is essential to derive adequate design criteria for these regions. The reader should consult the pertinent regional literature to best represent the expected reality under field conditions.

Advantages of the rule-of-thumb approach:

- It is very simple to use.

Disadvantages of the rule-of-thumb approach:

- It does not account for different water usage practices, pre-treatment technologies, climate, or influent wastewater concentrations.

- It does not account for non-ideal flow.
- It does not consider the geometry of the wetland cell or specific design approaches to minimise the risk of clogging.

Regression equations

Regression equations have also been used to design TWs. These equations are generated from a large collection of data. They generally require one or two input values (inlet concentration or mass load, and possibly HLR) and produce an estimate for expected effluent concentration. Note that the "goodness of fit" of the regression is sometimes quite poor. Table 2.8 provides some example regression equations for designing a HF wetland. An extensive list of regression equations for HF wetlands can be found in Rousseau *et al*. (2004).

Table 2.8 Example regression equations for HF wetlands.

Parameter	Equation [a, b]	Input Range [a,b]	Output Range [a,b]	R^2
BOD_5	$M_o = (0.13 \times M_i) + 0.27$	$6 < M_i < 76$	$0.32 < M_o < 21.7$	0.85
	$C_o = (0.11 \times C_i) + 1.87$	$1 < C_i < 330$	$1 < C_o < 50$	0.74
COD	$M_o = (0.17 \times M_i) + 5.78$	$15 < M_i < 180$	$3 < M_o < 41$	0.79
TSS	$M_o = (0.048 \times M_i) + 4.7$	$3 < M_i < 78$	$0.9 < M_o < 6.3$	0.42
	$C_o = (0.09 \times C_i) + 0.27$	$0 < C_i < 330$	$0 < C_o < 60$	0.67
TN	$M_o = (0.67 \times M_i) - 18.75$	$300 < M_i < 2{,}400$	$200 < M_o < 1{,}550$	0.96
TP	$M_o = (0.58 \times M_i) - 4.09$	$25 < M_i < 320$	$20 < M_o < 200$	0.61
	$C_o = (0.65 \times C_i) + 0.71$	$0.5 < C_i < 19$	$0.1 < C_o < 14$	0.75

[a] M_i and M_o are mass loads into and out of the system, respectively, in kg/ha·d (Vymazal, 1998).
[b] C_i and C_o are concentrations into and out of the system, respectively, in mg/L (Brix, 1994).

Advantages of using regression equations:

- They are simple to use.
- They take into account influent water quality (and sometimes HLR).
- They inherently account for background concentration (C^*) because equations were created from actual water quality data from full-scale systems.

Disadvantages of using regression equations:

- They are only applicable if the design of the new wetland falls within the data range from which the regression equations were created.
- Many regression equations were created from very large treatment wetland systems, and may not apply to smaller systems.
- Flow rate is not always considered.
- The wetland area cannot be determined from equations that only correlate concentration or mass.

Plug-flow k-*C**

The first-order plug-flow *k*-*C** approach takes into account influent and effluent concentrations as well as background concentration, but assumes ideal plug-flow hydraulics (see Section 2.3). This approach is currently less used by design engineers, but is still often reported in the literature. Equation 2.10, adapted to incorporate *C**, can be used to solve for the wetland area, *A*, as follows (Equation 2.18):

$$A = \frac{Q_i}{k_A} \ln\left(\frac{C_o - C^*}{C_i - C^*}\right) \quad (2.18)$$

where:

C_o = outlet concentration, mg/L

C_i = inlet concentration, mg/L

C^* = background concentration, mg/L

k_A = modified first-order areal rate coefficient, m/d

Q_i = influent flow rate, m^3/d

Additionally, Equation 2.14 (see Section 2.3) can be used to correct the reaction rate coefficient k_A to the anticipated climate conditions for the new wetland design.

Advantages of the plug-flow *k*-*C** approach:

- It takes into account influent concentration (C_i), background concentration (C^*), HLR (q) and areal reaction rate coefficient (k_A).
- It can take into account temperature correction factor (θ).

Disadvantages of the plug-flow *k*-*C** approach:

- It does not account for non-ideal flow, which creates a large risk, especially when low effluent concentrations must be achieved (Kadlec and Wallace, 2009).

- There is no guidance as to which k_A-value to choose (for example, when a range of reaction rate coefficients are reported).

The assumption of ideal plug-flow hydraulics has been widely reported in the literature as inaccurate (Kadlec, 2000), and thus is no longer recommended for use.

Mass loading charts

Another possible approach is the use of mass loading charts. The small-scale treatment wetland design manual written by Wallace and Knight (2006) was created from a collection of water quality data from over 1,500 small-scale treatment wetlands around the world. The data was used to create scatter plots that display influent mass loading rates versus effluent concentrations. This design manual is the first of its kind to consider the concept of risk tolerance in wetland design.

The loading charts in Wallace and Knight (2006) provide a visualisation of the risk tolerance of the design, including lines that correspond to the 50th, 75th, and 90th percentile of data collected (Figure 2.8). Using these charts, the design of a new small-scale treatment wetland can be chosen based on influent mass loading rate, desired effluent concentration, and risk tolerance. A design chosen based on the 50th percentile indicates that a system would meet the desired effluent concentration 50% of the time. A design chosen based on the 90th percentile line would be predicted to meet the desired effluent concentration 90% of the time (e.g., nine times out of ten), but would require a much larger area.

Advantages of the mass loading chart approach:

- It accounts for influent and effluent concentration (C_i and C_o), as well as inflow rate (Q_i).
- It inherently accounts for background concentration (C^*) and non-ideal flow because the charts were created from actual water quality data from full-scale systems.
- It allows the designer to choose the level of risk tolerance for a given design.

Disadvantages of the mass loading chart approach:

- It does not explicitly account for reaction rate coefficients (k_A or k_V) or temperature correction (θ).
- It does not explicitly consider the geometry of the wetland cell or specific design approaches to minimise the risk of clogging. This must be checked separately.

*P-k-C** approach

Wetland area A can be calculated by rearranging Equation 2.13:

$$A = \frac{PQ_i}{k_A}\left(\left(\frac{C_i - C^*}{C_o - C^*}\right)^{\frac{1}{P}} - 1\right) = \frac{PQ_i}{k_V h}\left(\left(\frac{C_i - C^*}{C_o - C^*}\right)^{\frac{1}{P}} - 1\right) \qquad (2.19)$$

where:

C_o = outlet concentration, mg/L
C_i = inlet concentration, mg/L
C^* = background concentration, mg/L
h = wetland water depth, m
k_A = first-order areal rate coefficient, m/d
k_V = first-order volumetric rate coefficient, 1/d
P = apparent number of tanks-in-series (TIS), dimensionless
Q_i = influent flow rate, m^3/d

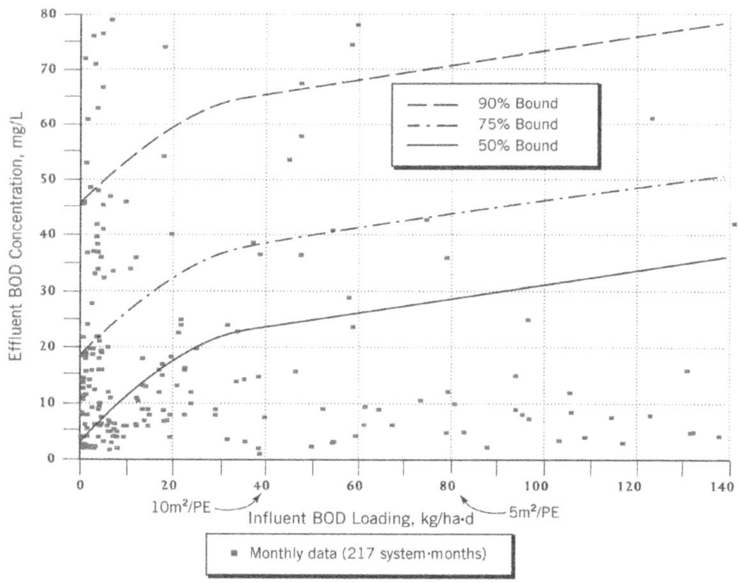

Figure 2.8 BOD$_5$ Loading chart for small-scale HF wetlands providing secondary treatment of domestic wastewater. Conversion to other loading units: 10 kg/ha·d = 1 g/m^2·d. Reprinted with permission from Wallace and Knight (2006).

Values for *P*, *k*, and *C** can be chosen using the information provided in Section 2.3 (Tables 2.6, 2.3, and 2.5, respectively), and the wetland area can be subsequently calculated. Temperature correction factors (θ-values, Table 2.4) can be used to adjust *k*-rates to the climate conditions of a specific location (if known). As with all design calculations, it is important to double-check that the appropriate units are used for each value chosen so that the result makes sense.

Advantages of the *P-k-C** approach include:

- It accounts for influent and effluent concentration (C_i and C_o), as well as background concentration (C^*).
- It accounts for areal or volumetric reaction rate coefficients (k_A or k_V) and temperature correction factors (θ).
- The designer can choose the level of risk (50%, 80% or 90% compliance) for certain design variables.

Disadvantages of the *P-k-C** approach include:

- There are many variables to assess and many have only limited information from which to select appropriate design values for a specific condition.
- The value of *P* depends on the geometry of the wetland unit, and its selection needs to take this into account.
- The designer must be extremely familiar with all of the material provided in Kadlec and Wallace (2009) in order to understand and locate the required design information.

2.5 ASSESSMENT OF TREATMENT PERFORMANCE

This section analyses how water losses may influence the assessment of removal efficiencies and effluent concentrations. It was seen in Section 2.2 (which discussed water balance in wetlands) that ET is one of the important components in this balance. When ET losses are higher than the gains from precipitation (assuming a wetland with bottom sealing and no infiltration to groundwater), the outflow will be lower than the inflow.

In dry periods with no important rainfall, ET may play an important role in the reduction of the effluent flow. ET is not simple to measure, but reported values may range between 0 and 50 mm/d (0 to 0.050 $m^3/m^2 \cdot d$), and research studies in Brazil indicated ET values in the order of 5 to 30 mm/d (0.005 to 0.030 $m^3/m^2 \cdot d$) for HF systems (Costa, 2013). Depending on the surface hydraulic rate applied (q or HLR_s) on the wetland unit, this loss may represent an important fraction. For instance, for q equal to 0.060 $m^3/m^2 \cdot d$ and an ET of 0.015 $m^3/m^2 \cdot d$, this means that 25% of the water is lost to the atmosphere, and the effluent flow will be only 75% of the influent one.

This of course affects the HRT in the wetland and, as mentioned in Section 2.2, the average flows between inlet Q_i and outlet Q_o could be used when computing retention time. However, it was emphasised that for the sake of simplicity, unless otherwise stated, only the inlet flow Q_i is considered in the computations shown in this volume.

Water losses have another implication. Water that is lost through ET is pure water (pollutant concentration equal to zero). This means that this water loss has the effect of increasing the outlet concentrations (mg/L). Traditionally, in large wastewater treatment, influent and effluent flow are assumed to be the same, simplifying efficiency calculations. However, because of ET losses in wetlands, the influent and effluent flow rates can be different and removal efficiencies should be calculated based on mass fluxes (Equation 2.20):

$$E = \frac{Q_i C_i - Q_o C_o}{Q_i C_i} \quad (2.20)$$

where:

E = removal efficiency

Q_i = influent flow rate, m^3/d

C_i = influent concentration, mg/L

Q_o = outlet flow rate, m^3/d

C_o = outlet concentration, mg/L

This provides for a more accurate representation of the actual removal efficiency of the treatment system. For instance, for a measured inlet BOD_5 concentration of C_i=200 mg/L and a measured outlet concentration of C_o=40 mg/L, the removal efficiency, based on the simplified calculation, would be (200-40)/200 = 0.80 = 80%. However, when flows are different in the influent and the effluent, if the wetland has 25% percent of water loss through ET, the removal efficiency based on loads would be [(1×200)-(1-0.25) × 40]/[1×200] = 0.85 = 85%, according to Equation 2.20. This is the actual removal that took place in the wetland unit. When specifying removal efficiencies in a report, the author must always specify how removal efficiency was computed. Within the same concept, the measured effluent concentration is a result of the actual removal that took place, but also of the increase in concentration due to the water losses. One way of correcting this, and obtaining the outlet concentration, independent of the ET, is given by:

$$\text{Corrected } C = \text{Measured } C\left(1 - \text{fraction of water loss}\right) \quad (2.21)$$

In the same example, the measured outlet concentration was 40 mg/L. The wetland had 25% of water loss. The corrected concentration, resulting from removal mechanisms only, is 40 × (1-0.25) = 30 mg/L. If there were no water losses, this would be the measured effluent concentration, but because 25% of the water was lost, the outlet was more concentrated and was measured as 40 mg/L. In this volume, unless otherwise stated, the outlet concentrations are reported as the measured ones. Indeed, reaction coefficients (k_A and k_V) calculated based on measured influent and effluent field data already incorporate the effects of removal mechanisms and the influence of water losses.

3

Horizontal flow wetlands

3.1 INTRODUCTION AND APPLICATION

The HF wetland configuration originated from the pioneering work in Germany in the late 1960s. In the literature, the abbreviation HSSF is also commonly used, standing for horizontal subsurface flow constructed wetlands. Whilst the design has evolved to typically rely on gravel or coarse sand rather than clay-rich soils, the concept of passing wastewater horizontally through a porous medium remains the same.

Horizontal flow wetlands are used for secondary and tertiary treatment of domestic wastewater, as well as for a variety of industrial effluents (Vymazal and Kröpfelová, 2008; Kadlec and Wallace, 2009). For HF wetlands treating domestic wastewater, primary treatment is generally achieved via a septic tank or an Imhoff tank. These systems are widely used in the Czech Republic, Spain, Portugal, Nicaragua, and North America among other countries for secondary treatment of domestic wastewater (Vymazal and Kröpfelová, 2008). In warm climate regions, it is common to find HF wetlands following septic tanks, anaerobic baffled reactors (ABR) and UASB reactors. In the UK, HF wetlands are predominantly used for tertiary treatment, with over 600 HF wetlands in operation (CWA Database, 2011). In this scenario, secondary treatment is often achieved using biological treatment units such as rotating biological contactors or trickling filters, and the HF wetlands are used as a polishing step. Additionally, combinations of HF with other wetland types (VF, FWS) have been used in a variety of hybrid systems.

In a typical HF wetland, the gravel bed is saturated and planted with emergent wetland plants (Figure 3.1). Water enters the treatment system at one end, flows through the gravel media, and is collected on the opposite end of the bed prior to being discharged. A standpipe located outside of the wetland bed controls the water level within the gravel media. The whole bed is isolated from the surrounding land by a combination of a plastic liner and a geotextile membrane.

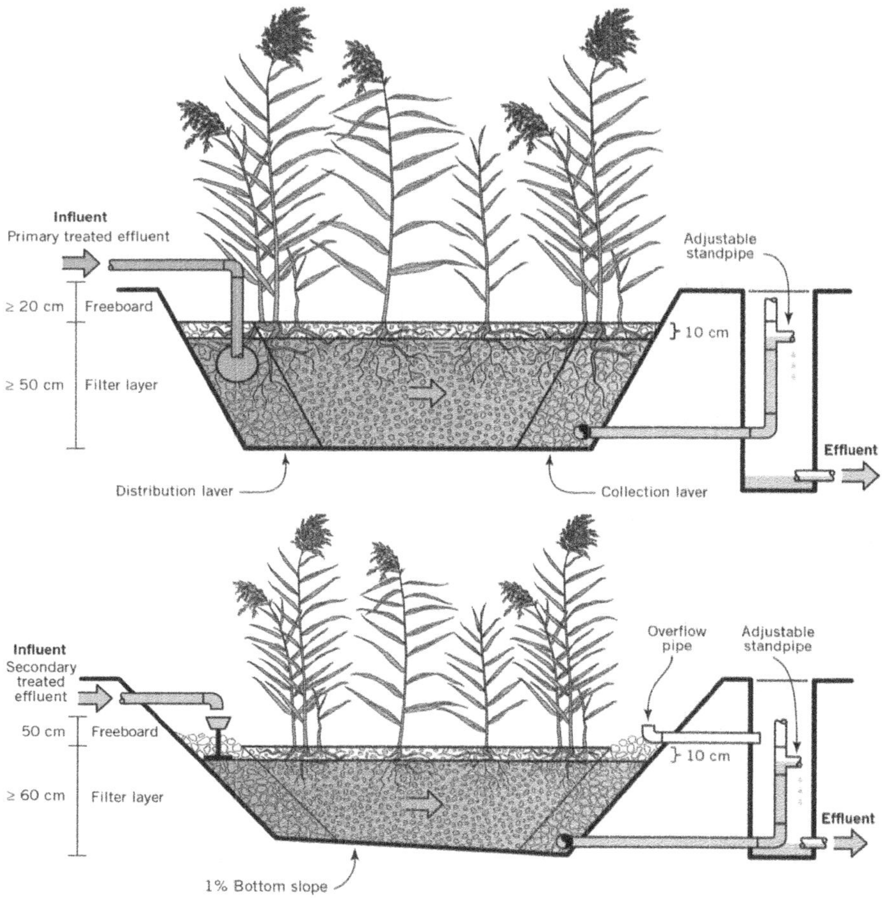

Figure 3.1 Typical schematic of a HF wetland; top: secondary treatment; bottom: tertiary treatment of domestic wastewater.

For secondary treatment of domestic wastewater, the gravel depth is generally 0.5 to 0.7 m and the water level is kept 5 – 10 cm below the surface. In tertiary treatment applications in the UK, the depth of the basin itself is 1.0 to 1.5 m, of which approximately 0.60 m is filled with gravel. HF systems in the UK are generally constructed with a longitudinal sloped base (1%) to facilitate draining of the bed if needed. The remaining bed volume is used for water storage during high flows or storm events.

3.2 DESIGN AND WATER QUALITY TARGETS

The predominant microbiological removal pathways in HF wetlands are anaerobic. When used for secondary treatment of domestic wastewater, HF are generally capable of removing BOD_5 and TSS to a reasonable extent (20 mg/L in the effluent) but the performance of individual systems depends heavily on influent concentrations and HLRs. Removal of TN in HF systems is somewhat restricted due to limited aerobic conditions for nitrification. However, HF wetlands can be very effective at denitrification provided that there is sufficient nitrate and carbon present in the water column. Phosphorus is not sustainably removed in HF wetlands over the long term unless reactive media is used (see Section 6.2).

Design guidance for HF wetlands varies greatly. They can be sized using simple specific surface area requirements (m^2/PE), maximum areal loading rates (for example, g $BOD_5/m^2 \cdot d$), or more sophisticated methods such as loading charts or the *P-k-C** approach (Section 2.3). Table 3.1 summarises the main design parameters of HF wetlands providing secondary or tertiary treatment of domestic wastewater for select countries. In general, the design criteria for HF wetlands providing secondary treatment of domestic wastewater are quite similar between different countries within the same climatic conditions.

Length-to-width ratios for secondary HF wetlands generally fall between 2:1 and 4:1, whereas for tertiary systems width is typically greater than the length to maximise the cross-sectional area and reduce clogging potential with the higher hydraulic rates applied. Some practitioners also apply a higher width to length ratio even in secondary systems, to try to minimise clogging on highly loaded systems. Most design guidelines specify a maximum loading rate based on the wetland plan area, as this is simple to explain to builders and end users. The underlying assumption is that all HF beds provide a standard depth of 0.6 m of media – a legacy of the earlier beliefs that the plant roots provided the majority of treatment and this value being the assumed maximum root depth penetration. The use of a maximum cross-sectional area loading, i.e., the load applied at the inlet width and depth, moves away from this assumption and provides opportunity to modify bed length and depth to enable sustainable treatment of the wastewater. The bed width, however, is typically limited to a maximum of 25 – 30 m to facilitate even flow distribution into a single wetland cell.

The distribution and collection of wastewater is of critical importance to ensure the pollutants come into contact with the microorganisms whilst minimising bed clogging. The beds typically have a coarser media at both ends (Figure 3.2). Outlet collection systems are typically agricultural drainage pipes, with holes or slots, positioned across the width of the wetland bed end, connected to a swivel pipe to control water depth within the bed (Figure 3.2). Subsurface loading structures are typically pipes with tees or orifices evenly spaced every 10% of the bed width (Vymazal and Kröpfelová, 2008), whereas surface loading structures are typically troughs with v-notches spaced at 2.5 m intervals (Griffin *et al.*, 2008). In the past, riser pipes were used but as experience showed these were difficult to keep clean, they were replaced by open troughs that can easily be cleaned with a suction hose or shovel.

Table 3.1 Main design parameters of HF wetlands for select countries.

	Czech Republic	**Spain**	**US**	**UK**
Treatment Step	Secondary	Secondary	Secondary	Tertiary
Pre-treatment	Screens + Imhoff tank	Screens + septic tank	Septic tank	Primary settling + biological treatment
Specific surface area requirement (m^2/PE)	5	10	5 – 10	0.7
Maximum areal organic loading rate (g $BOD_5/m^2 \cdot d$)	–	6	4 – 8	2 – 13
Maximum cross-sectional organic loading rate (g $BOD_5/m^2 \cdot d$)	–	–	250 [a]	–
Hydraulic loading rate (mm/d)	–	20	20 – 40	200
Gravel size (mm)	< 20	5 – 6	> 4	10 – 12
Distribution system	Subsurface pipes	Subsurface pipes	Subsurface pipes	Surface trough
Reference	Vymazal (1996) Vymazal and Kröpfelová (2008)	García and Corzo (2008)	Wallace and Knight (2006)	Cooper *et al.* (1996) Griffin *et al.* (2008)

[a] This value has been reduced to 100 g $BOD_5/m^2 \cdot d$ in a recent proposal by Wallace (2014).

Figure 3.2 Examples of civil structures in HF wetlands; left: distribution troughs; right: water level control structure.

There is variability in the grain size of media used depending on the country and the designer's preference (Table 3.1). The effect of media size on wetland sizing is considered when performing hydraulics calculations to prevent overland flow and is reflected on the maximum recommended loading rates for each design variation.

In Europe, HF wetlands are typically planted with common reed (*Phragmites* sp.). The systems can be planted with other types of plants, depending on local regulations and/or climate. For example, in the United States, plants from the *Phragmites* genus are considered an invasive species, so other species such as *Sagittaria latifolia, Schoenoplectus validus, Schoenoplectus acutus* and *Iris pseudacorus* are used (Wallace and Knight, 2006). In tropical climates, plants such as *Cyperus, Typha, Helicornia* and *Canna* sp. have been used (Rani *et al.*, 2011).

The role of plants in HF wetlands is mainly related to physical processes such as providing increased surface area for attached microbial growth, and for providing better filtration of TSS. In temperate and cold climates, the litter layer can provide extra thermal insulation during the winter. However, in hot, arid climates, it may be necessary to cut the vegetation on a regular (annual) basis. This is because the climatic conditions favour net accumulation of litter, needlessly insulating the bed whilst reducing the wetland storage capacity. For HF wetlands providing secondary treatment of domestic wastewater, the contribution of plant uptake to nutrient removal is minimal. Plant-mediated oxygen transfer occurs, but is minimal in comparison to the oxygen demand exerted by the incoming wastewater (Brix, 1990; Tanner and Kadlec, 2003).

3.3 OPERATION AND MAINTENANCE

No treatment wetland system is maintenance free. The most critical operational issue for HF wetlands is clogging. This occurs when the pore spaces in the media are filled with solids (organic or inorganic), instead of wastewater, thus limiting the contact area and time between the biofilm and the water. Clogging can occur in any kind of (biological) filter and has been reported for both HF and VF systems (Knowles *et al.*, 2011). For HF wetlands providing treatment of domestic wastewater, clogging is most commonly caused by excessive organic and/or solids loading onto the gravel bed. This is often due to improper maintenance of the septic tank (secondary treatment HF wetland) or final settling tanks (tertiary HF wetlands), or poor dimensioning of the wetland itself. Hydraulic and solids loading rates that are at the top end of recommended values have been suggested as the main factors resulting in the reported clogging of HF systems. This can be a result of inadequate design or of a deliberate use of HF beds for solids storage rather than treatment (Dotro and Chazarenc, 2014). In either case, it is the net accumulation of solids in the pore spaces that results in overland flow and a clogged system. Clogging can thus be minimised and the bed life extended by selecting appropriate media (e.g., gravel vs. sand) and loading rates (checking both hydraulic and mass pollutant loads) as explained in Chapter 2, and ensuring the upstream processes are correctly maintained to enable the bed to operate within the range of its intended design.

Routine checks for proper O&M of HF wetlands include:

- Upstream treatment: Septic tanks (secondary treatment HF) and final settling tanks (tertiary treatment HF) must be emptied regularly to prevent solids carryover to the HF wetland. The emptying interval depends on the size of the septic tank, but should be conducted at least once per year. A similar logic applies to settling tanks after other forms of biological treatment ahead of a tertiary HF, with typical emptying frequencies for rotating biological contactors and trickling filters anywhere between 30 and 90 days. Sludge that has been removed from the tanks can be treated onsite in a separate sludge treatment wetland, or transported to a centralised treatment plant for further processing. In addition, if pumping is required, the equipment must be maintained according to the manufacturer's specifications (e.g., lubrication).

- Influent distribution system: Uneven distribution can result in a solids or organic loading over a small portion of the intended influent area, and result in clogging. For surface-loaded systems, it is important to ensure that wastewater is evenly delivered across the width of the wetland bed. For HF wetlands that have subsurface loading, the distribution pipes must be properly designed and should contain inspection ports so that the influent header can be periodically washed out and/or cleaned.

- Outlet control structure: The outlet level control structure should be checked on a routine basis. The water level should be maintained 5 – 10 cm below the surface of the gravel. If a decrease in the height of the outlet control structure does not result in a decrease in the water level within the gravel, further investigations may be necessary to assess the extent of clogging in the gravel bed.

- Surface sludge accumulation (surface-loaded tertiary HF wetlands only): Surface-loaded tertiary treatment systems should be monitored for sludge accumulation. Sludge accumulation at the inlet zone of the bed should be measured once a year. In the O&M plan, action limits should be set to trigger intervention (refurbishment) actions based on the sludge build-up rate (cm/yr) and the available storage capacity within the freeboard of the wetland cells.

- Vegetation: Wetland vegetation should be monitored to ensure that unwanted plant species (weeds) do not overtake the intended plant community. In the first two full growing seasons, weeds should be removed as needed. In temperate climates, the plant litter provides extra insulation during the winter. In hot and arid climates, thatch may accumulate indefinitely and plant harvesting may be necessary.

Problems that arise from issues during design, construction, and operation of HF wetlands include:

- Improperly maintained pre-treatment: Solids carryover from improperly maintained upstream treatment components could result in the release of solids to the gravel bed and result in clogging.

- Unsuitable filter media: Only washed, rounded gravel or coarse clean sand should be used in HF wetlands. Unwashed media can contain a high content of fines, which can lead to clogging. Sharp edges can damage the liner and provide less ideal pore spaces thus affecting porosity.

- Uneven distribution of influent wastewater: Uneven distribution of wastewater along the wetland width can lead to localised clogging and preferential flow paths.

- Poor site grading and/or lack of berms: Rainwater flow into the HF bed can become problematic if the wetland is not built with berms or the site is not adequately graded to divert runoff away from, rather than into, the wetland basin.

In situations where a HF wetland has become clogged to the degree that water is short-circuiting over the surface of the bed and thus bypassing treatment, or in the case that effluent discharge requirements are no longer being met, refurbishment of the bed may be necessary. Refurbishment, which has most often been implemented in the UK, includes complete removal of the gravel media. Partial replacement of the media is not recommended because wastewater will preferentially flow through the clean media. The gravel can either be removed and disposed off-site, or washed onsite and returned to the bed (Murphy *et al.*, 2009). Other methods for alleviating clogged areas of wetlands include the use of earthworms (Davison *et al.*, 2005; Li *et al.*, 2011) or the injection of an oxidising agent such as hydrogen peroxide into the gravel bed (Nivala and Rousseau, 2009). However, to date, these alternative methods have only been reported for a handful of full-scale systems.

3.4 DESIGN EXAMPLE – ONSITE SYSTEM

Design a HF wetland for a single-family home (5 PE) in a temperate climate. BOD_5 effluent target is 30 mg/L.

Assumptions:

- A septic tank for pre-treatment, and that the septic tank removes 1/3 of the BOD_5 load.
- An average per capita wastewater generation of 150 L/d and a per capita BOD_5 load of 60 g per person and day (DWA, 2017).

Perform the design according to the methods described in Chapter 2:

- Rule-of-thumb
- Regression equation
- Plug-flow *k-C**
- Mass loading charts
- *P-k-C**

Summary of inputs to the HF wetland:

$$\text{Inflow, } Q_i = 5\ \text{PE} \times 150 \frac{\text{L}}{\text{PE} \cdot \text{d}} \times \frac{1\ \text{m}^3}{1000\ \text{L}} = 0.75\ \frac{\text{m}^3}{\text{d}}$$

$$\text{Mass Load In, } M_i = 5\ \text{PE} \times 60 \frac{\text{g BOD}}{\text{PE} \cdot \text{d}} \times \frac{2}{3} = 200 \frac{\text{g BOD}}{\text{d}}$$

$$\text{Concentration In, } C_i = \frac{M_i}{Q_i} = 200 \frac{\text{g BOD}}{\text{d}} \div 0.75\ \frac{\text{m}^3}{\text{d}} = 266 \frac{\text{mg BOD}}{\text{L}}$$

Rule-of-thumb

Choose a rule-of-thumb guideline. For example, according to the Danish guideline (Brix and Johansen, 2004) HF wetlands are sized at 5 m^2/PE.

$A = 5\ \text{PE} \times 5\ \text{m}^2/\text{PE} = 25\ \text{m}^2$

According to the Danish guideline, HF wetlands sized at 5 m^2/PE are expected to achieve 90% reduction in BOD_5, which should result in an effluent concentration close to 25 mg/L (0.1 × 266 mg/L = 27 mg/L). Note that any further increase in the influent water quality (e.g., septic tank effluent) would result in an increase in expected effluent BOD_5 concentrations.

A length-to-width ratio between 2:1 and 4:1 is common for HF wetlands. Choosing a length-to-width ratio of three yields the following calculation:

Knowing:

$$A = l \times w$$

$$\frac{l}{w} = 3$$

Solve for w:

$$w = \sqrt{\frac{A}{3}} = \sqrt{\frac{25}{3}} = 2.9 \text{ m}$$

Choosing a length-to-width ratio of three results in a wetland that is 2.9 m wide by 8.7 m long (total area of 25.2 m^2).

These dimensions, although exact, are not practical to use in the field. Engineering designs must take into consideration the constructability of the system. Small treatment wetland systems, especially those built for individual homes, are often constructed by homeowners themselves or small contractors. Choosing wetland dimensions that are easy to measure and implement in the field is an important aspect of the design process. Choosing a wetland of 3.0 m wide by 8.5 m long (total area of 25.5 m^2) results in a length-to-width ratio of 2:8, and system dimensions that are much easier to measure and implement during construction. When adjusting the width of the wetland bed, it is generally better to increase the width rather than to decrease it. Decreasing the width will increase the overall cross-sectional organic loading rate and increase chances of clogging. Typical saturated depth for a HF wetland treating septic tank effluent is 0.5 m.

Regression equations

Example regression equations are given in Table 2.8. An example regression equation for BOD_5 removal in HF wetlands is:

$C_o = (0.11 \times C_i) + 1.87$ (with the constraints $1 < C_i < 330$ and $1 < C_o < 50$).

Expected BOD_5 effluent concentration is then:

$$(0.11 \times 266 \text{ mg/L}) + 1.87 = 31 \text{ mg/L}$$

Such a system, in principle, should produce an effluent BOD_5 concentration slightly higher than 30 mg/L, but the calculation does not produce a recommended area for the wetland.

Plug-flow k-C*

Step 1. Select k-rate

Locate the appropriate values for k_A:
$k_A = 25$ m/yr (Table 2.3 for BOD_5 and 50th percentile)

Step 2. Check input parameters and unit conversions. Calculate minimum required area

As with any engineering equation, it is extremely important to check that the unit labels are consistent. Failing to convert Q_i (which is often given in L/d) and k_A (which is often given in m/yr) into compatible units will yield incorrect calculations.

The following values are converted to correct units (where necessary) and Equation 2.21 can be used:

$$\text{Inflow, } Q_i = 5\ \text{PE} \times 150\frac{\text{L}}{\text{PE}\cdot\text{d}} \times \frac{1\ \text{m}^3}{1000\ \text{L}} = 0.75\frac{\text{m}^3}{\text{d}} \times \frac{365\ \text{d}}{1\ \text{yr}} = 273.75\ \frac{\text{m}^3}{\text{yr}}$$

$$\text{Concentration In, } C_i = \frac{M_i}{Q_i} = \frac{200\frac{\text{g BOD}}{\text{d}}}{0.75\frac{\text{m}^3}{\text{d}}} = 266\frac{\text{mg BOD}}{\text{L}}$$

$$\text{Concentration Out, } C_o = 30\frac{\text{mg}}{\text{L}}\quad (\textit{given})$$

$$\text{Background Concentration, } C^* = 10\frac{\text{mg}}{\text{L}}\quad \text{(Table 2.5)}$$

$$\text{First-order areal rate coefficient}, k_A = 25\frac{\text{m}}{\text{yr}}$$

$$A = -\frac{Q_i}{k_A}\ln\left(\frac{C_o - C^*}{C_i - C^*}\right) = -\frac{273.75\frac{\text{m}^3}{\text{yr}}}{25\frac{\text{m}}{\text{yr}}}\ln\left(\frac{30\frac{\text{mg}}{\text{L}} - 10\frac{\text{mg}}{\text{L}}}{266\frac{\text{mg}}{\text{L}} - 10\frac{\text{mg}}{\text{L}}}\right) = 27.9\ \text{m}^2$$

Step 3. Choose wetland dimensions

A length-to-width ratio between 2:1 and 4:1 is common for HF wetlands. Choosing a length-to-width ratio of three yields the following calculation:

Knowing:

$$A = l \times w$$

$$\frac{l}{w} = 3$$

Solve for w:

$$w = \sqrt{\frac{A}{3}} = \sqrt{\frac{27.9}{3}} = 3.0 \text{ m}$$

Choosing a length-to-width ratio of three results in a wetland that is 3 m wide by 9.3 m long (total area of 27.9 m^2). These dimensions, although exact, are not practical to use in the field.

Choosing a wetland of 3.0 m wide by 10.0 m long (total area of 30 m^2) results in a length-to-width ratio of 3.3, and system dimensions that are much easier to measure and implement during construction. Typical saturated depth for a HF wetland treating septic tank effluent is 0.5 m.

Step 4. Check cross-sectional organic loading rate

Clogging is a commonly cited problem in HF wetlands, and can occur when large length-to-width ratios are chosen. Wallace and Knight (2006) recommend a maximum cross-sectional organic loading rate of 250 g BOD_5/m^2·d.

The cross-sectional area of the wetland is:

$$3 \text{ m} \times 0.5 \text{ m} = 1.5 \text{ m}^2$$

The influent BOD_5 loading to the wetland is 200 g/d (calculated earlier).

The cross-sectional organic loading is therefore:

$$200 \ \frac{\text{g BOD}}{\text{d}} \div 1.5 \text{ m}^2 = 133 \ \frac{\text{g BOD}}{\text{m}^2 \cdot \text{d}}$$

This is well below the recommended maximum of 250 g/m^2·d, so the wetland is not likely to clog on a medium term.

Mass loading charts

Loading charts can give an indication for effluent concentration based on influent mass loading.

Step 1. Choose desired confidence interval

Choose desired confidence interval on chart (50%, 75% or 90%). An example chart is shown in Figure 2.8. A full collection of charts is provided in Wallace and Knight (2006). Locate corresponding desired effluent concentration and corresponding influent mass loading rate.

With mass loading charts, choosing a confidence interval of 50% indicates that five of every ten effluent samples will have an effluent concentration below the desired outlet concentration, as long as the wetland being sized has similar wastewater characteristics and is located in similar (temperate) climate conditions. New designs with particularly strong or weak wastewater, or those located in extreme climates, should not be made using the mass loading chart approach.

In this example, assuming similar wastewater and climate conditions, the 50% bound line crosses 30 mg/L effluent concentration at an influent loading of approximately 90 kg/ha·d (refer to Figure 2.8).

Step 2. Calculate required wetland area

Areal mass loading rate (from mass loading chart)

$$90\frac{\text{kg}}{\text{ha}\cdot\text{d}}\times\frac{1000\text{ g}}{1\text{ kg}}\times\frac{1\text{ ha}}{10{,}000\text{ m}^2}=9\frac{\text{g}}{\text{m}^2\cdot\text{d}}$$

Mass BOD_5 load into the wetland (from assumptions) = 200 g BOD_5/d

Total daily BOD_5 load divided by the mass load of BOD_5 into wetland equals the wetland area:

$$A=\frac{200\frac{\text{g BOD}}{\text{d}}}{9\frac{\text{g}}{\text{m}^2\cdot\text{d}}}=22.2\text{ m}^2$$

Note that the determination of wetland area is highly dependent on the desired effluent concentration. If an effluent concentration of 20 mg/L BOD_5 is desired, the required area (based on the information in the provided chart) would be approximately 80 m^2. Note that HF wetlands treating septic tank effluent generally do not produce very low effluent concentrations, which is why the confidence intervals of 75% and 90% in this example either result in a very large calculated area or cannot be used for design purposes.

Step 3. Choose wetland dimensions

A length-to-width ratio between 2:1 and 4:1 is common for HF wetlands. Choosing a length-to-width ratio of three yields the following calculation:

Knowing:

$$A = l \times w$$

$$\frac{l}{w} = 3$$

Solve for w:

$$w = \sqrt{\frac{A}{3}} = \sqrt{\frac{22.2}{3}} = 2.7 \text{ m}$$

Choosing a length-to-width ratio of three results in a wetland that is 2.7 m wide by 8.1 m long (total area of 21.9 m^2).

Choosing a wetland of 3.0 m wide by 8.0 m long (total area of 24 m^2) results in a length-to-width ratio of 2.7, and system dimensions that are much easier to measure and implement during construction. Typical saturated depth for a HF wetland treating septic tank effluent is 0.5 m.

Step 4. Check cross-sectional organic loading rate

Clogging is a commonly cited problem in HF wetlands, and can occur when large length-to-width ratios are chosen. Wallace and Knight (2006) recommend a maximum cross-sectional organic loading rate of 250 g $BOD_5/m^2 \cdot d$ (Table 3.1).

The cross-sectional area of the wetland is:

$$3.0 \text{ m} \times 0.5 \text{ m} = 1.5 \text{ m}^2$$

The influent BOD_5 loading to the wetland is 200 g/d (calculated earlier).

The cross-sectional organic loading is therefore:

$$200 \frac{\text{g BOD}}{\text{d}} \div 1.5 \text{ m}^2 = 133 \frac{\text{g BOD}}{\text{m}^2 \cdot \text{d}}$$

This is well below the recommended maximum of 250 g $BOD_5/m^2 \cdot d$, so the wetland is not likely to clog on the medium term.

*P-k-C** approach

A currently suggested approach for treatment wetland sizing is the *P-k-C** modified first-order equation with a non-zero background concentration.

Step 1. Select k-rate

Locate the appropriate values for k_A:
k_A = 25 m/yr. (Table 2.3 and 50th percentile)

Step 2. Check input parameters and unit conversions. Calculate minimum required area

As with any engineering equation, it is extremely important to check that the unit labels are consistent. Failing to convert Q (which is often given in L/d) and k_A (which is often given in m/yr) into compatible units will yield incorrect calculations.

The following values are converted to correct units (where necessary) and plugged into Equation 2.22:

$$C_i = 266\frac{\text{mg}}{\text{L}} \text{ (calculated earlier)}$$

$$C_o = 30\frac{\text{mg}}{\text{L}} \text{ (given)}$$

$$C^* = 10\frac{\text{mg}}{\text{L}} \text{ (Table 2.6)}$$

$$P = 3 \text{ (Table 2.7)}$$

$$Q_i = 273.75\ \frac{\text{m}^3}{\text{yr}} \text{ (calculated earlier)}$$

$$k_A = 25\frac{\text{m}}{\text{yr}}$$

$$A = \frac{PQ_i}{k_A}\left(\left(\frac{C_i - C^*}{C_o - C^*}\right)^{\frac{1}{P}} - 1\right) = \frac{3 \times 273.75\ \frac{\text{m}^3}{\text{yr}}}{25\ \frac{\text{m}}{\text{yr}}}\left(\left(\frac{266\ \frac{\text{mg}}{\text{L}} - 10\ \frac{\text{mg}}{\text{L}}}{30\ \frac{\text{mg}}{\text{L}} - 10\ \frac{\text{mg}}{\text{L}}}\right)^{\frac{1}{3}} - 1\right) = 44.0\ \text{m}^2$$

Step 3. Choose wetland dimensions

A length-to-width ratio between 2:1 and 4:1 is common for HF wetlands. Choosing a length-to-width ratio of three yields the following calculation:

Knowing:

$$A = l \times w$$

$$\frac{l}{w} = 3$$

Solve for w:

$$w = \sqrt{\frac{A}{3}} = \sqrt{\frac{44.0}{3}} = 3.8\ \text{m}$$

A length-to-width ratio of three results in a wetland that is 3.9 m wide by 11.5 m long (total area 44 m^2). These dimensions, although exact, are not practical to use in the field. As in the previous examples, it is better to choose dimensions that are easy to implement during construction.

Choosing a wetland of 4.0 m wide by 11.0 m long (total area of 44 m^2) results in a length-to-width ratio of 2.75. Typical saturated depth for a HF wetland treating septic tank effluent is 0.5 m.

Step 4. Check cross-sectional organic loading rate

Clogging is a commonly cited problem in HF wetlands, and can occur when length-to-width ratios that are too large are chosen. Wallace and Knight (2006) recommend a maximum cross-sectional organic loading rate of 250 $g/m^2 \cdot d$.

The cross-sectional area of the wetland is:

$$4\ \text{m} \times 0.5\ \text{m} = 2\ \text{m}^2$$

The influent BOD_5 loading to the wetland is 200 g/d (calculated earlier).

The cross-sectional organic loading is therefore:

$$200\ \frac{\text{g BOD}}{\text{d}} \div 2\ \text{m}^2 = 100\ \frac{\text{g BOD}}{\text{m}^2 \cdot \text{d}}$$

This is well below the recommended maximum of 250 $g/m^2 \cdot d$, so the wetland is not likely to clog in the medium term.

Summary

Each design approach provides a different result for the HF wetland area (Table 3.2):

Table 3.2 Summary of calculated HF wetland area for a single-family home (5 PE) in a temperate climate.

Method	Will the wetland produce an effluent concentration of 30 mg/L?	Minimum calculated wetland area (m^2)	Minimum practical wetland area (m^2)
Rule-of-thumb	Yes	25.0	25.5
Regression equation	No	–	–
Plug-flow *k-C**	Yes	27.9	30.0
Mass loading chart	Yes	22.2	24.0
*P-k-C**	Yes	44.0	44.0

Note that the use of regression equations cannot always provide enough information for wetland sizing. The rule-of-thumb approach is arguably the easiest to use, but care must be taken that the new design falls within the assumptions that were used to create the sizing recommendation. The plug-flow *k-C** approach, which is often reported in the literature, is no longer recommended for use in design. In this example, the Mass Loading Chart provides the least conservative result, which is half the area from the *P-k-C** approach. For small-scale systems, especially at the household level, a slightly oversized system will be able to better cope with fluctuations in influent flow and load. However, as the number of homes increases, the fluctuations in flow and load will decrease, and oversizing a system can inflate construction costs to the point where a treatment wetland is no longer a cost-effective treatment option. The *P-k-C** approach provides a design that explicitly accounts for the most up-to-date information on removal rate coefficients, wetland hydraulics and pollutant weathering, as well as background concentrations. However, like all other design approaches, it is only valid for the climatic conditions where it has been developed. Most design information available has been developed in temperate climates, and cannot be applied one-to-one in warm climate regions.

In this example, the only target pollutant was BOD_5 and is thus the simplest scenario for design. As mentioned in Chapter 2, in practice, most treatment systems have multiple water quality targets (e.g., BOD_5, TSS, TN). In such cases, the calculations must be repeated for each pollutant. The limiting factor will result in the largest system footprint, and this value should be selected for design to ensure the treatment system meets all water quality targets.

3.5 DESIGN EXAMPLE – COMMUNITY

Design a HF wetland for a small community (100 PE) in a warm climate.

Assumptions:

- BOD_5 effluent target is 30 mg/L.
- An UASB reactor is used for pre-treatment, and the UASB reactor removes two-thirds of the BOD_5 load (~ 67%, typical removal efficiency of UASB reactors – see von Sperling, 2007a).
- Because the UASB reactor provides a biological treatment (even though not a very efficient one), the wetland will receive an influent from a secondary treatment.
- Assume an average per capita wastewater generation of 120 L/d and a per capita BOD_5 load of 50 g per person and day.

Perform the design according to the *P-k-C** method, which is currently the preferred procedure. The other design methods have already been illustrated in the previous example (Section 3.4).

The design of the UASB reactor is not shown here. The design of this unit is fully described and exemplified in Volume 4 (Anaerobic Reactors) of this series (Chernicharo, 2007).

Summary of wastewater generated (influent to UASB reactor):

$$\text{Inflow, } Q_i = 100\ \text{PE} \times 120 \frac{\text{L}}{\text{PE}\cdot\text{d}} \times \frac{1\ \text{m}^3}{1000\ \text{L}} = 12\ \frac{\text{m}^3}{\text{d}}$$

$$\text{Mass Load In, } M_i = 100\ \text{PE} \times 50 \frac{\text{g BOD}}{\text{PE}\cdot\text{d}} = 5000 \frac{\text{g BOD}}{\text{d}}$$

$$\text{Concentration In, } C_i = \frac{M_i}{Q_i} = 5000 \frac{\text{g BOD}}{\text{d}} \div 12\ \frac{\text{m}^3}{\text{d}} = 417 \frac{\text{g BOD}}{\text{m}^3} = 417 \frac{\text{mg BOD}}{\text{L}}$$

Summary of inputs to the HF wetland (effluent from UASB reactor):

$$\text{Inflow, } Q_i = 12\ \frac{\text{m}^3}{\text{d}}$$

$$\text{Mass Load In, } M_i = 5000 \frac{\text{g BOD}}{\text{d}} \times \left(1 - \frac{2}{3}\right) = 1667 \frac{\text{g BOD}}{\text{d}}$$

$$\text{Concentration In, } C_i = \frac{M_i}{Q_i} = 1667 \frac{\text{g BOD}}{\text{d}} \div 12\ \frac{\text{m}^3}{\text{d}} = 139 \frac{\text{g BOD}}{\text{m}^3} = 139 \frac{\text{mg BOD}}{\text{L}}$$

Step 1. Select k-rate

Locate the appropriate values for k_A:

k_A value for treatment of primary effluents is 25 m/yr and for secondary effluents is 37 m/yr (see Table 2.3). Because the treatment provided by the UASB reactor is secondary, but not very efficient, an intermediate value of k_A = 32 m/yr (20°C) will be adopted in this design.

Step 2. Check input parameters and unit conversions. Calculate minimum required area

As with any engineering equation, it is extremely important to check that the unit labels are consistent. Failing to convert *Q* (which is often given in L/d) and k_A (which is often given in m/yr) into compatible units will yield incorrect calculations. The following values are converted to correct units (where necessary) and inserted into Equation 2.22.

$$C_i = 139\frac{\text{mg}}{\text{L}} \text{ (calculated earlier)}$$

$$C_o = 30\frac{\text{mg}}{\text{L}} \text{ (given)}$$

$$C^* = 7\frac{\text{mg}}{\text{L}} \text{ (value in between 5 } \frac{\text{mg}}{\text{L}} \text{ [treatment of secondary effluent]}$$

$$\text{and 10 } \frac{\text{mg}}{\text{L}} \text{ [treatment of primary effluent], see text following Table 2.5)}$$

$$P = 3 \text{ (Table 2.6)}$$

$$Q_i = 12\ \frac{\text{m}^3}{\text{d}} \times 365\frac{d}{yr} = 4380\ \frac{\text{m}^3}{yr} \text{ (calculated earlier)}$$

$$k_A = 32\frac{\text{m}}{\text{yr}} \text{ (calculated earlier)}$$

$$A = \frac{PQ_i}{k_A}\left(\left(\frac{C_i - C^*}{C_o - C^*}\right)^{\frac{1}{P}} - 1\right) = \frac{3 \times 4380\ \frac{\text{m}^3}{\text{yr}}}{32\ \frac{\text{m}}{\text{yr}}}\left(\left(\frac{139\frac{\text{mg}}{\text{L}} - 7\frac{\text{mg}}{\text{L}}}{30\frac{\text{mg}}{\text{L}} - 7\frac{\text{mg}}{\text{L}}}\right)^{\frac{1}{3}} - 1\right) = 325\ \text{m}^2$$

Step 3. Choose wetland dimensions

In order to give operational flexibility, two wetlands in parallel will be adopted (n=2). Therefore, the area of each unit will be:

$$A_1 = A_2 = \frac{A_{Total}}{2} = \frac{325\ \text{m}^2}{2} = 162\ \text{m}^2$$

A length-to-width ratio between 2:1 and 4:1 is common for HF wetlands. Choosing a length-to-width ratio of l/w=3 yields the following calculation:

Knowing:

$$A = l \times w$$

$$\frac{l}{w} = 3$$

Solve for w:

$$w = \sqrt{\frac{A}{3}} = \sqrt{\frac{162}{3}} = 7.3\ \text{m}$$

Therefore, the length is $l = 3 \times w = 3 \times 7.3\ \text{m} = 21.9\ \text{m}$.

Adopting round values of length l=22.0 m and width w=8.0 m will lead to an area of 176 m^2 per unit, and a total area of 2x176 m^2 = 352 m^2.

This value of 352 m^2, for a population of 100 PE, corresponds to a net per capita land requirement of 352/100 = 3.5 m^2/PE.

The depth for the liquid will be adopted as h=0.5 m, which is a typical saturated depth for a HF wetland.

With these dimensions, the saturated volume (media and liquid) will be:

- Each unit: $V_1 = V_2 = l \times w \times h = 22\ \text{m} \times 8\ \text{m} \times 0.5\ \text{m} = 88\ \text{m}^3$
- Total: $V_{Total} = V_1 + V_2 = 2 \times 88\ \text{m}^3 = 176\ \text{m}^3$

The total media bed volume is comprised by the saturated volume plus the height above the liquid level. Adopting an additional unsaturated media depth of 0.10 m will lead to a total bed height of 0.5+0.1=0.6 m. The total media bed volume of both units will be $2 \times (22\ \text{m} \times 8\ \text{m} \times 0.6\ \text{m}) = 211\ \text{m}^3$.

Step 4. Check HRT and surface loading rates

Assuming a porosity of ε = 0.35, the theoretical HRT will be (Equation 2.9):

$$\tau = \frac{V \times \varepsilon}{Q_i} = \frac{176\ \text{m}^3 \times 0.35}{12\ \text{m}^3/d} = 5.1\ \text{d}$$

The resulting surface HLR q will be (Equation 2.10):

$$q = \frac{Q_i}{A} = \frac{12\ \text{m}^3/\text{d}}{352\ \text{m}^2} = 0.034 \frac{\text{m}^3}{\text{m}^2 \cdot \text{d}} = 34\ \frac{\text{mm}}{\text{d}}$$

The surface organic loading rate will be:

$$\frac{M_i}{A} = \frac{1667\ \text{g BOD/d}}{352\ \text{m}^2} = 4.7 \frac{\text{g BOD}}{\text{m}^2 \cdot \text{d}}$$

All the loading values are within reasonable values, according to the literature. However, it is known that in several warm-climate locations, HF wetlands perform well with higher loading rates compared with temperate climates, that is, with smaller volumes and areas. The designer may consider that these calculations with the *P-k-C** method have been too conservative for the climatic conditions under study, since most experience with its use lies in temperate climates.

To gain more insight and verify the resulting surface area, the design based on the mass loading charts may be used, as it was shown to be the least conservative of the approaches. From Figure 2.8, for an effluent BOD_5 of 30 mg/L, the recommended organic loading rate for a 50th percentile is 90 kg BOD_5/(ha·d) or 9.0 g $BOD_5/(m^2 \cdot d)$. This is almost double the loading rate that resulted from the *P-k-C** method (4.7 g $BOD_5/(m^2 \cdot d)$), implying that the surface area could be halved. On the other hand, if the 75th percentile is considered in the design, the loading rate would be substantially smaller than the one calculated from the *P-k-C** model at 1.5 g $BOD_5/(m^2 \cdot d)$. Coming back to the 50th percentile, the loading rate calculated from the *P-k-C** model of 4.7 g $BOD_5/(m^2 \cdot d)$ corresponds to an effluent BOD_5 concentration of 25 mg/L. Small differences in the desired effluent quality may have a considerable impact on the required surface area.

This uncertainty is typical in design. The designer must reflect on these implications and decide on the adoption of an approach that delivers the right balance of safety and feasibility. This is also an incentive to the development of design guidelines that cater for the specific conditions of warm climates. For the sake of this example, the dimensions calculated with the *P-k-C** method will be used.

Step 5. Check cross-sectional organic loading rate

Clogging is a commonly cited problem in HF wetlands, and can occur when length-to-width ratios that are too large are chosen. Wallace and Knight (2006) recommend a maximum cross-sectional organic loading rate of 250 g $BOD_5/(m^2 \cdot d)$.

The cross-sectional area of each wetland unit is:

$$8 \text{ m} \times 0.5 \text{ m} = 4 \text{ m}^2$$

The influent BOD_5 loading to the wetland system is 1667 g BOD_5/d (calculated earlier), or, for each unit, $M_i/n = 1.667/2 = 834$ g BOD_5/d per wetland.

The cross-sectional organic loading is therefore:

$$834 \ \frac{\text{g BOD}}{\text{d}} \div 4 \text{ m}^2 = 209 \ \frac{\text{g BOD}}{\text{m}^2 \cdot \text{d}}$$

This is below the recommended maximum of 250 g $BOD_5/(m^2 \cdot d)$, so the wetland is not likely to clog in the medium term.

Step 6. Specify other dimensions and details in the wetland units

In the inlet and outlet zones, a buffer zone with larger stones to allow better distribution of the influent and collection of the effluent will be provided. Usual values are between 0.5 and 1.0 m of length. In the present design, a value of 0.7 m will be adopted. The size of the stones in these inlet and outlet zones may be between 10 and 20 cm.

The additional bed volume associated with these zones is:

- Inlet zone: $V = l \times w \times h = 0.7 \text{ m} \times 8 \text{ m} \times 0.6 \text{ m} = 3.4 \text{ m}^3$
- Outlet zone: $V = l \times w \times h = 0.7 \text{ m} \times 8 \text{ m} \times 0.6 \text{ m} = 3.4 \text{ m}^3$

The grain size in the filter bed varies, according to different design criteria (see Table 3.1). In the present case, in order to further reduce risks of clogging, a diameter in the upper bound of the values presented in Table 3.1 will be adopted: effective diameter d_{10} = 16 mm.

The dimensions of the basin must be such as to accommodate the bed and a freeboard above the top surface of the bed. In this example, a value of 0.2 m is adopted. Therefore, the total height of the basin is 0.8 m, of which 0.6 m is for the bed (being 0.5 m saturated and 0.1 m non-saturated) and 0.2 m is for the freeboard.

The dimensions calculated are for rectangular longitudinal and cross sections. The wetland cells may be built with sloped banks with compacted soil of good quality, to facilitate construction. In the case of units with sloped banks, the dimensions provided before are for the bottom of the saturated bed. Note that some designers prefer to use the top of the gravel bed for the length specification, as this is what will be visible after the wetland is built. However, this will mean the treatment area and treatment volume for the system will be smaller than the design specifications, as the bottom of the bed will effectively be shorter than the calculated value. For this reason, some designers (and what is recommended in this volume) apply the calculated length for the treatment area at the bottom of the cell, resulting in a longer bed when looking at the plan area once the system is built. For the specification of bank slopes and the calculation of the dimensions for an inverted pyramid trunk, the reader is referred to Chapter 9 of Volume 3 (Stabilisation Ponds) of this series of books (von Sperling, 2007c).

A longitudinal slope of the bottom level, between 0.5 and 1.0%, towards the outlet end may be adopted. This can be used to facilitate emptying of the bed, but requires additional work. In this example, no slope of the bottom was included. The wetland would be planted with a species well-adapted to the climatic conditions of the region. This can be checked in various literature sources from academic journals or books (e.g., Kadlec and Wallace, 2009). For simplicity, *Typha sp.* is selected here as it is widespread in the warm climates of South America.

Step 7. Make schematic drawings of the system

The schematic arrangement of the system in plan view is presented in Figure 3.3, and a schematic longitudinal section for one unit is presented in Figure 3.4. Both drawings are not to scale. The dimensions length l=22 m and width w=8 m are measured at the bottom of the wetlands (conservative approach). With the side-slope this then gives 23.6 m × 9.6 m as measured on the top.

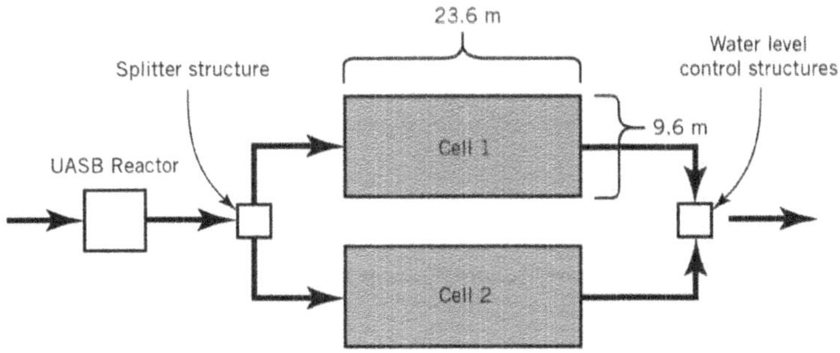

Figure 3.3 Schematic arrangement of the system (not to scale).

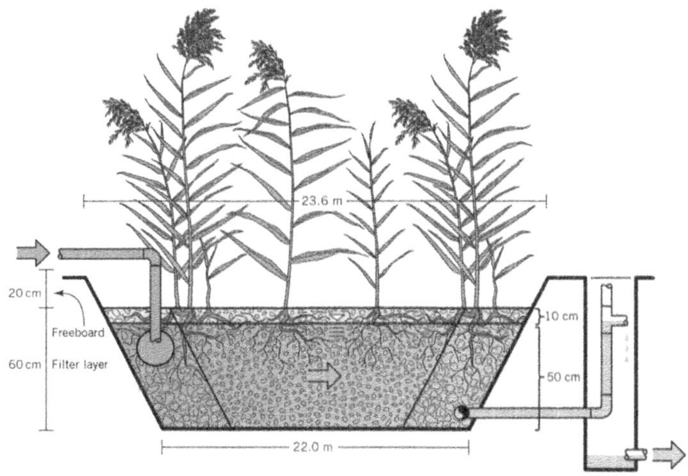

Figure 3.4 Longitudinal section of one unit (not to scale).

3.6 CASE STUDY

The case study presents the oldest HF wetland in the Czech Republic (Vymazal, 2009). The HF wetland in Ondřejov was built in 1991 and has been in operation since then. The wetland treats sewage from 360 persons in a single HF bed planted with common reed (*Phragmites australis*). The system consists of a grit chamber, an Imhoff tank and a single 806 m^2 HF bed filled with gravel (3 – 15 mm) (Figures 3.5 and 3.6). The average measured flow was 56.3 m^3/d resulting in a HLR of 70 mm/d (Vymazal, 2009).

Table 3.3 shows performance data of the HF wetland in Ondřejov for the period 1991 to 2015. The average yearly removal efficiencies for this period have been 93% for BOD_5, 89% for COD, 95% for TSS, 41% for NH_4-N, 37% for TN (data until 2004 only), and

46% for TP. Similar performance has been reported for other HF wetlands in the Czech Republic (Vymazal, 2011) and other countries (Vymazal and Kröpfelová, 2008).

Figure 3.5 Ondřejov HF wetland: sand trap (front), Imhof tank and vegetated HF bed (back). Photo courtesy of Jan Vymazal.

Figure 3.6 Ondřejov HF wetland: HF bed with vegetation and inlet distribution zone. Photo courtesy of Jan Vymazal.

Table 3.3 Performance data for the HF wetland in Ondřejov for the period 1991-2015 (values are annual averages, data from 1991 to 2004 from Vymazal, 2009; data from 2005 onwards provided by Jan Vymazal; "-" means that the parameter was not measured in that year).

Year	BOD_5		COD		TSS		NH_4-N		TN		TP	
	In (mg/L)	Out (mg/L)	In (mg/L)	Out (mg/L)	In (mg/L)	Out (mg/L)	In (mg/L)	Out (mg/L)	In (mg/L)	Out (mg/L)	In (mg/L)	Out (mg/L)
1991	168	16	660	62	108	5	56.5	2.5	-	-	14.2	2.5
1992	161	38	254	74	44	10	19.6	17.8	-	-	17.7	21.1
1993	152	28	445	88	257	18	29.8	22.6	41.8	28.8	6.4	5.8
1994	330	16	719	57	304	3	39.1	21.6	49.8	28.0	12.6	6.8
1995	83	13	188	35	67	5	17.7	17.0	34.6	21.1	4.2	4.8
1996	106	13	207	44	36	9	-	10.7	26.1	15.4	-	1.9
1997	112	13	-	45	-	23	-	23.0	57.2	27.0	-	6.0
1998	137	12	-	33	-	4	-	25.5	36.3	28.8	-	5.4
1999	493	27	-	79	-	7	-	40.0	86.9	49.0	-	8.1
2000	571	30	-	68	-	8	-	41.6	81.5	44.1	-	10.1
2002	169	15	190	36	66	2	40.3	33.1	57.1	34.1	8.3	7.0
2003	408	11	1,047	40	463	9	39.6	20.8	58.2	28.5	24.0	7.7
2004	187	16	458	81	148	6	33.0	17.0	20.6	21.6	34.0	4.5
2005	204	16	558	51	515	20	33.6	23.1	-	-	6.0	3.4
2006	340	3	540	33	290	11	33.5	7.9	-	-	11.0	4.7
2007	360	8	580	53	310	13	13.7	8.5	-	-	5.8	1.6
2008	195	9	370	35	190	8	29.0	10.0	-	-	6.0	1.3
2009	220	12	450	58	200	10	54.0	10.0	-	-	7.0	3.0
2010	421	14	1,032	64	710	6	17.0	15.5	-	-	7.1	3.2
2011	257	7	1,060	41	355	8	50.0	28.0	-	-	7.2	3.2
2012	355	4	1,210	29	596	3	38.3	24.0	-	-	6.5	3.5
2013	255	6	984	33	417	3	25.5	14.4	-	-	6.6	2.4
2014	305	11	974	26	818	6	33.1	23.5	-	-	5.0	3.9
2015	395	15	695	78	458	8	47.7	30.8	-	-	7.7	3.5
Average	**266**	**15**	**631**	**52**	**318**	**8**	**34.3**	**20.4**	**50.0**	**29.7**	**10.4**	**5.2**

4

Vertical flow wetlands

4.1 INTRODUCTION AND APPLICATION

The standard VF wetland gained prominence in the 1990s in response to changing legal requirements in Europe (specifically in Austria, Denmark, and Germany) that required elimination of ammonia nitrogen for small wastewater treatment plants. Their main application is for secondary treatment of domestic wastewater. A large number of variants of VF wetlands exist (Stefanakis *et al.*, 2014), including French VF wetlands treating raw wastewater (Chapter 5) and sludge TWs (Section 8.3). Additionally, combinations of VF with other wetland types (HF, FWS) have been used in a variety of hybrid systems.

Figure 4.1 shows a schematic of a typical VF wetland in Europe. The sand and/or gravel bed is planted with emergent macrophytes. Primary treated wastewater is loaded intermittently to the filter surface, and the large amount of water from a single loading causes good distribution of inflow water on the surface. The water percolates through the substrate then gradually drains and is collected by a drainage network at the base of the filter. Between loadings, oxygen re-enters the pore space of the media, transporting oxygen into the filter bed in order to sustain aerobic microbial processes. The whole bed is isolated from the surrounding land by a combination of a plastic liner and a geotextile membrane.

In other countries in which legal regulations do not allow spreading of wastewater on the surface due to potential contact with humans (e.g. USA) and in regions with very cold climates, VF wetland influent distribution systems are insulated with a layer of gravel or mulch. In cold climates where freezing is a concern, the distribution system must also drain completely between doses.

Due to the highly oxidising conditions in the filter bed, VF wetlands with intermittent loading are extremely efficient for removal of organic carbon (BOD_5 or COD). They are also suitable when strictly aerobic processes such as nitrification are required (Langergraber and Haberl, 2001; Garcia *et al.*, 2010). Thus, VF wetlands are commonly used for secondary as well as tertiary treatment of domestic wastewater. VF wetlands are also used to treat landfill leachate and food processing wastewaters, which often contain high levels of ammonium nitrogen and/or organic carbon (upwards of hundreds of milligrams per litre) (Kadlec and Wallace, 2009), as well as other agro-industrial wastewaters such as olive mill effluents, dairy farm wastewater and animal farm effluent (Stefanakis *et al.*, 2014).

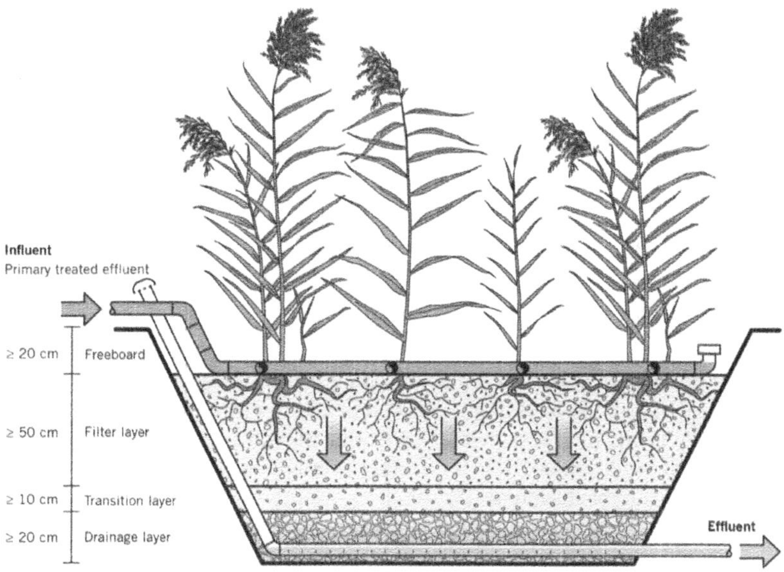

Figure 4.1 Schematic of a typical VF wetland in Europe.

As emphasised in this volume (mainly in Chapter 2), the unsaturated conditions of the filter medium imply that several hydraulic considerations, modelling approaches and design specifications that are exclusive for saturated media (HF wetlands) are not applicable here. The reader is referred to these observations in the preceding text in order not to use inadequate concepts for the design of intermittently loaded VF wetlands.

4.2 DESIGN AND WATER QUALITY TARGETS

In general, contaminants that are degraded aerobically are easily removed using VF wetlands with intermittent loading. For domestic and municipal wastewater, organic matter (BOD_5 or COD) and ammonia nitrogen are removed mainly through aerobic microbial processes. Solids (such as TSS) and pathogenic organisms are removed by physical filtration. The treatment efficiency of a VF wetland is directly related to the filter material used. If fine material is used, the retention time of the wastewater in the filter is longer, often enabling higher removal efficiencies; however, the HLRs are limited, as it takes longer for water to infiltrate and the potential for clogging increases. Coarser filter material enables higher HLRs and less clogging potential, but results in lower removal efficiencies. This can be partially overcome in some cases by increasing the depth of the main layer. Available design guidelines for VF wetlands are based on empirical rules-of-thumb, such as those using specific surface area requirements (Brix and Johansen, 2004; DWA, 2017; ÖNORM, 2009).

Table 4.1 summarises the main design parameters of VF wetlands in Denmark, Germany and Austria. In general, the design parameters are similar in these countries. These guidelines provide design advice for achieving legal requirements. The Danish guideline states that, when VF wetlands are designed according to the guideline, they will remove 95% of BOD_5 and 90% of ammonia nitrogen and thus will meet the legal requirements (e.g., effluent concentrations below 10 mg BOD_5/L and 5 mg NH_4-N/L, respectively). In Austrian and German legislation, nitrification is not required in winter, i.e. the effluent standard of 10 mg NH_4-N/L must be met only if the effluent wastewater temperature is above 12°C (valid in Austria for plants smaller than 500 PE and in Germany for plants smaller than 1,000 PE; for larger plants more stringent regulations apply).

Besides the parameters given in Table 4.1, all guidelines require a drainage layer of gravel at the bottom of the bed and an intermediate or transition layer (e.g. 10 cm gravel of 4 – 8 mm in diameter) between main and drainage layer. The intermediate layer prevents grains from the filtration layer from migrating into the drainage layer. The coarse gravel in the drainage layer allows for good drainage and together with the drainage pipes, provides oxygen to the deepest layer of the bed. In order to avoid migration of fine gravel to the coarser gravel layers below, the Terzaghi rule of $D_{15}/d_{85} \leq 4$ is used (D corresponds to the transition layer and d to the main layer) (Sherard *et al.*, 1984).

The design guidelines include a non-compulsory top layer of gravel (e.g. 4 – 8 mm) to prevent erosion during intermittent loading as well as to allow no free water on the surface. Additionally, an additional top layer increases thermal insulation and ensures higher temperatures of the filter in winter (about 1 – 2°C for a 15 cm top layer). However, the main disadvantage is that the top layer reduces oxygen supply to the main layer and fixes the stems of emergent macrophytes so that they cannot move and break up the surface of the main layer in the non-loading periods. Both effects lead to less degradation of particulate organic matter at the surface of the main layer and thus to a higher risk of clogging. If a top layer is used, it should be limited to a depth of 5 – 10 cm (Langergraber *et al.*, 2009a).

Emergent macrophytes, most often *Phragmites australis* (common reed), are used for planting VF wetlands. The roles of the macrophytes in relation to pollutant removal in VF wetlands are mainly related to physical processes. The roots provide surface area for attached microbial growth, and root growth is known to help maintain the hydraulic properties of the filter. The vegetation cover protects the surface from erosion. In temperate climates, litter provides an insulation layer on the wetland surface for operation during winter. Uptake of nutrients plays a minor role for common wastewater parameters compared to the degradation processes caused by microorganisms. If wetland plants are not harvested, some nutrients absorbed by the plant will be released to the system during decomposition, resulting in a possible secondary pollutant release to the wetland. Some plants also release organic compounds, which can be used to aid in denitrification. Compared to the amount of oxygen brought into the system from the atmosphere due to intermittent loading, release of oxygen through roots plays a minor role in VF wetlands (Brix, 1997).

Table 4.1 Main design parameters of VF wetlands treating primary treated domestic wastewater.

Design Parameter	Denmark [a]	Germany	Austria
Minimum size	5 PE	4 PE	4 PE
Primary treatment (septic tank)	2 m^3 for a single household (5 PE)	0.3 m^3/PE (min. 3 m^3)	0.25 m^3/PE (min. 2 m^3)
Specific surface area requirement (m^2/PE)	3	4	4
Max. organic loading rate (g COD/$m^2 \cdot$d)	27	20	20
Main layer			
Filter material	Sand	Sand 0.06 – 2 mm	Sand 0.06 – 4 mm
Depth (cm)	100	> 50	> 50
d_{10} (mm)	0.25 – 1.2	0.2 – 0.4	0.2 – 0.4
d_{60} (mm)	1 – 4	-	-
$U = d_{60}/d_{10}$	< 3.5	< 5	-
Distribution System	-	Minimum one opening hole per m^2	Minimum one opening hole per 2 m^2
Reference	Brix and Johansen (2004) Brix and Arias (2005)	DWA (2017)	ÖNORM B 2505 (2009)

[a] for VF wetlands up to 30 PE, the Danish guidelines require recirculation of 50% of the effluent to the 1st chamber of the septic tank.

Intermittent loading of the VF wetlands is achieved with a pump or, if the landscape allows and adequate slope is available, intermitted dosing can be achieved with siphons (which do not require external energy). In any case, a good distribution of the wastewater on the surface of the VF wetland must be guaranteed to utilise the whole filter volume. The maximum HLR should not exceed 80 L/m²·d (0.08 m³/m²·d) (DWA, 2017), the interval between loadings should be ≥ 6 hours (DWA, 2017), or between 3 and 6 hours (ÖNORM, 2009).

The distribution pipes should have a diameter of about 40 mm with circular opening holes with a diameter of not less than 8 mm to avoid blocking of the openings with solids. For single-stage VF wetlands, according to German and Austrian design guidelines, opening holes have to be distributed evenly on the surface of the filter with minimum one opening hole per m² surface area. To avoid freezing in the distribution pipes, it must be guaranteed that no water is accumulating in the pipes after a single loading (DWA, 2017; ÖNORM, 2009).

For larger VF wetlands (> 100 PE), it is common practice to divide the surface area into several smaller ones that can be loaded independently, thereby allowing one of these areas to be in resting phase and not loaded. The German design guidelines (DWA, 2017) recommend VF wetlands should be designed so that a quarter of the total surface is in resting phase. The Austrian design guidelines (ÖNORM, 2009) recommend for a VF bed a maximum surface area of 400 m² to guarantee even distribution of the wastewater on the surface.

The design guidelines have been developed in temperate climates in which operation in winter is the critical operational condition. The slower degradation processes in winter limit the amount of organic matter that can be loaded to VF wetlands without clogging. However, in warmer climates and when systems are operated seasonally (i.e. only in summer), the specific surface area requirements are lower. Experience from Austria shows that the same treatment efficiency can be reached with VF wetlands designed with 2 m²/PE if operated only in summer, meaning effluent water temperatures ≥ 10°C (Langergraber *et al.*, 2007). Experiences on the use of VF wetlands in warmer climates exist for several regions. Besides a reduced specific surface area, hydraulic loads up to 200 mm/d can be applied in warm climate regions (Hoffmann *et al.*, 2011; Stefanakis *et al.*, 2014).

An alternative design model for VF wetlands that is based on oxygen demand was proposed by Platzer (1999). This model is based on the oxygen requirements for aerobic processes (oxidation of COD and nitrification). The dimensioning criterion is that oxygen input (OI) needs to be greater than the oxygen demand (OD):

$$OI[g/d] - OD[g/d] > 0 \tag{4.1}$$

$$OD[g/d] = OD_{COD} + OD_{Nitrification} - OD_{Denitrification} \tag{4.2}$$

$$OI[g/d] = OI_{\text{Diffusion}} + OI_{\text{Convection}} \tag{4.3}$$

where:

OI = total oxygen input, g/d
OD = total oxygen demand, g/d
OD_{COD} = oxygen demand due to decomposition of organic matter, g/d
$OD_{\text{Nitrification}}$ = oxygen demand due to nitrification, g/d
$OD_{\text{Denitrification}}$ = decreased oxygen demand due to partial denitrification, g/d
$OI_{\text{Diffusion}}$ = oxygen input due to diffusion, g/d
$OI_{\text{Convection}}$ = oxygen input due to convection, g/d

According to Platzer (1999) the oxygen demand due to decomposition of organic matter is calculated with 0.7 g O_2/g COD and an average removal efficiency of 85% for COD in VF wetlands. Oxygen demand for nitrification is calculated with 4.3 g O_2/g Total Kjeldahl Nitrogen (TKN) and the oxygen saved due to denitrification is 2.9 g O_2/g COD. Assuming complete nitrification and about 10% denitrification in a VF wetland the oxygen demands can be calculated as follows:

$$OD_{\text{COD}} = 0.85 \times 0.7 \frac{\text{g O}_2}{\text{g COD}} \times COD_{\text{IN}} \tag{4.4}$$

$$OD_{\text{Nitrification}} = 4.3 \frac{\text{g O}_2}{\text{g TKN}} \times TKN_{\text{IN}} \tag{4.5}$$

$$OD_{\text{Denitrification}} = 0.10 \times 2.9 \frac{\text{g O}_2}{\text{g TKN}} \times TKN_{\text{IN}} \tag{4.6}$$

where:

OD_{COD} = oxygen demand due to decomposition of organic matter, g/d
$OD_{\text{Nitrification}}$ = oxygen demand due to nitrification, g/d
$OD_{\text{Denitrification}}$ = decreased oxygen demand due to partial denitrification, g/d
COD_{IN} = COD influent load, g/d
TKN_{IN} = TKN influent load, g/d

Oxygen input by diffusion is calculated with 1 g O_2/h·m^2. The time of saturation after loading has to be subtracted from the time between two loads. Platzer (1999) showed that oxygen input by diffusion is absent for 1.5 hours after each loading. Oxygen input by diffusion can thus be calculated by

$$OI_{\text{Diffusion}} = 1\left[\frac{\text{g O}_2}{\text{h}\times\text{m}^2}\right]\times A\ [\text{m}^2]\times\frac{24\ [\text{h}]-1.5\ [\text{h}]\times n_{\text{Loadings}}}{1\ [\text{d}]} \tag{4.7}$$

where:

$OI_{\text{Diffusion}}$ = oxygen input due to diffusion, g/d

A = surface area, m^2

n_{Loadings} = loadings per day, unitless

For calculating oxygen input by convection Platzer (1999) showed that convection is related to the hydraulic load. As air contains 300 mg O_2/L the oxygen input can be calculated as:

$$OI_{\text{Convection}} = 0.3\left[\frac{\text{g O}_2}{\text{L}}\right]\times V_{\text{Loaded}}\left[\frac{\text{m}^3}{\text{d}}\right]\times 1000\left[\frac{\text{L}}{\text{m}^3}\right] \tag{4.8}$$

where:

$OI_{\text{Convection}}$ = oxygen input due to convection, g/d

V_{Loaded} = volume of wastewater loaded per day, m^3/d

The design model based on oxygen demand was applied in South America where long-term experience now exists (Platzer *et al.*, 2016).

In a single stage VF wetland, removal of TN is limited to about 20 – 30%. If further nitrogen removal is required, two approaches have been shown to be successful:

1. Recirculation (see also Section 6.3): Effluent of the VF wetland is recirculated into the primary treatment and thus nitrate-rich water is brought into contact with organic matter. This aids in additional denitrification. Removal of TN is dependent on the recirculation rate. The increased hydraulic load generally increases the system size (when designed for recirculation) and creates additional costs for pumping.

2. Use of coarser filter material for the main layer combined with saturation of the drainage layer (see also Section 6.4): When using a coarse sand as filter material for the main layer (e.g. 1 – 4 mm or 2 – 3 mm) and a portion of the drainage layer is impounded, a TN removal rate of 60 – 70% can be reached without recirculation (Langergraber *et al.*, 2011).

Besides secondary treatment, VF wetlands are also used as a tertiary treatment stage. For tertiary treatment VF wetlands, the main design criteria in temperate climates is a maximum organic loading rate of 20 g COD/m^2·d. Compared to VF wetlands for secondary treatment, the HLRs in tertiary treatment VF wetlands can be higher and thus the specific surface area requirement can be reduced. The German design guidelines (DWA, 2017) recommend that the HLR of tertiary treatment VF wetlands should not exceed 120 L/m²·d (0.120 $m^3/m^2 \cdot d$) and that the interval between loadings should not be less than 3 hours.

4.3 OPERATION AND MAINTENANCE

The main operational problem of VF wetlands is clogging due to the insufficient removal of sludge from the primary treatment step (e.g. septic tank). If sludge is not removed, it will be transported to the filter surface and clog the filter. Several other operational problems can result from poor design and/or problems during the construction phase. Problems during design and/or construction that should be avoided include (adapted from Mitterer-Reichmann, 2012):

- Insufficient protection of VF wetland surface from surface water and superficial runoff: Soil substrate from the surrounding area is washed on the filter surface during rain events and causes clogging of the gravel and sand layers. To prevent this, border strips should be established around the filter beds.
- Unsuitable filter media: For economic and sustainability reasons, it is intended to use sand and gravel from as near as possible to the implementation site. When new providers are used, the grain size of the sand should be tested. The main problem that can occur is that too fine grain size distribution and/or unwashed sand or gravel that contains a large portion of fines can lead to clogging of the filter.
- Uneven slope of the filter surface: Ponding of water in single areas of the filter bed might lead to clogging.
- Intermittent loading system: Uneven distribution of wastewater causes uneven loading on parts of the VF wetland and can result in ponding (and eventual clogging). Thus, distribution pipes and opening holes must be evenly distributed over the surface of the filter bed and even distribution of wastewater must be ensured. Additionally, it is essential that the pipes drain completely after a loading event. Drilling a downward facing hole in the distribution system can facilitate this.
- Primary treatment using a septic tank: Poor quality concrete tanks can result in corrosion and sludge drift. In some cases, weathering of septic tank walls can occur. In the case that there is not enough ventilation into the tank, the cover of the septic tank should be perforated or air circulation achieved by other means.

Requirements for regular O&M of VF wetlands include (adapted from Mitterer-Reichmann, 2012):

- Maintenance logs: System owners should check nitrification of the VF wetland by measuring effluent ammonia nitrogen using a test kit on a monthly basis. The measurement should be recorded in a "maintenance book" together with all maintenance work done and operational problems that occur.

- Primary treatment: The sludge from the primary treatment unit must be removed in order to prevent sludge drift to the VF beds. The emptying interval depends on the volume of the tank, but sludge should be removed at least once a year. The sludge can be stabilised in a separate sludge treatment wetland onsite, or transported to a centralised wastewater treatment plant for further treatment.
- Intermittent loading: The intermittent loading can be checked by measuring the height difference in the well before and after a loading event.
- Siphons: After some years, the rubber part of some siphons can get porous, which allows wastewater to seep continuously and thus only one part of the VF filter is loaded. If this is not detected, the filter will become clogged after some time. Additionally, siphon hoses can split. Thus, the loading device should be checked once a month.
- Distribution pipes: In order to prevent freezing of wastewater in the pipes of the distribution system, it is essential that after a loading no water stays in the pipes. This needs to be checked at least in fall and after removing wetland plants.
- Wetland plants: During the first year, weeds should be removed until a mature cover of wetland vegetation is established. Wetland plants should be cut every two to three years either in spring or in fall. If cut in fall, the plant material should be left on the filter surface to provide an insulation layer.

4.4 DESIGN EXAMPLE

The following simple example shows the design of a VF wetland in a temperate climate. It is estimated that the VF wetland shall treat household wastewater of a small settlement with 50 PE, the average flow is 150 L/PE·d. Pollutant per capita generation rates of 60 g BOD_5, 120 g COD and 11 g TKN per capita and day are assumed (DWA, 2017; ÖNORM, 2009).

Step 1. Define influent flow and pollutant concentrations

$$\text{Inflow, } Q_i = 50\ \text{PE} \times 150 \frac{\text{L}}{\text{PE}\cdot\text{d}} \times \frac{1\ \text{m}^3}{1000\ \text{L}} = 7.5\ \frac{\text{m}^3}{\text{d}}$$

$$\text{COD Concentration (raw wastewater)} = 120\ \frac{\text{g COD}}{\text{PE}\cdot\text{d}} \div 150\ \frac{\text{L}}{\text{PE}\cdot\text{d}} \times \frac{1\ \text{mg}}{1000\ \text{g}} = 800\ \frac{\text{mg COD}}{\text{L}}$$

$$\text{TKN Concentration (raw wastewater)} = 11\ \frac{\text{g TKN}}{\text{PE}\cdot\text{d}} \div 150\ \frac{\text{L}}{\text{PE}\cdot\text{d}} \times \frac{1\ \text{mg}}{1000\ \text{g}} = 73.3\ \frac{\text{mg TKN}}{\text{L}}$$

Step 2. Design the three-chamber septic tank

According to ÖNORM B 2505 (2009) for populations greater than 50 PE, septic tanks are sized at 0.25 m^3/PE, with a minimum per capita tank surface area of 0.06 m^3/PE. The first chamber of the tank should comprise of 50% of the septic tank volume, whereas the second and third chambers should each be comprised of 25% of the total tank volume. The third chamber of the septic tank functions as the dosing tank for the wetland. It is assumed that one-third of the COD is removed in the septic tank (DWA, 2017).

Calculate minimum tank volume: $50 \text{ PE} \times 0.25 \dfrac{\text{m}^3}{\text{PE}} = 12.5 \text{ m}^3$

Selected septic tank volume = 13 m^3

Calculate minimum surface area of septic tank: $50 \text{ PE} \times 0.06 \dfrac{\text{m}^2}{\text{PE}} = 3 \text{ m}^2$

Selected surface area of septic tank = 3.5 m^2

Calculate depth of septic tank: $\dfrac{13 \text{ m}^3}{3.5 \text{ m}^2} = 3.7 \text{ m}$

Hydraulic retention time in first and second septic tank chamber: $\dfrac{13 \text{ m}^3 \times 0.75}{7.5 \dfrac{\text{m}^3}{\text{d}}} = 1.3 \text{ d}$

COD Concentration (post-septic tank) = $800 \dfrac{\text{mg COD}}{\text{L}} \times \dfrac{2}{3} = 533 \dfrac{\text{mg COD}}{\text{L}}$

Step 3. Design the VF wetland

Two different design approaches are shown. The first uses sand with a grain size of 0.06 – 4 mm (d_{10} = 0.3 mm; according to ÖNORM, 2009). The second approach uses coarse sand with grain size 2 – 3 mm (such as the first stage according to Langergraber *et al.*, 2011). The key parameters of the two designs are shown in Table 4.2.

Summary

In general, the coarser the filter media used for the main layer of the VF bed:

- The higher the acceptable hydraulic and organic loads.
- The smaller the required surface area of the VF bed.
- The smaller the single loading of the VF bed.
- The more opening holes are required to achieve good distribution of the wastewater on the surface.
- However, less pollutant removal (TSS, BOD_5, COD and NH_4-N) can be expected.

Table 4.2 Key design parameters and expected effluent concentrations of two different VF wetlands.

Main layer	**Sand (0.06 – 4 mm)**	**Coarse sand (2 – 3 mm)**
Surface area		
Maximum areal organic loading rate (g COD/m²·d)	20	80
Organic load (g COD/d) [a]	4,000	4,000
Required surface area (m²) [b]	200	50
Wetland cell configuration (m)	10 x 20	5 x 10
Intermittent loading		
Loading interval (hours)	6	2
Volume of a single dose (m³) [c]	1.875	0.625
Surface area of intermittent loading tank (m²) [d]	0.875	0.875
Height difference in intermittent loading tank (m) [e]	2.2	0.70
Distribution pipes		
Minimum one opening per	2 m²	1 m²
Expected effluent concentrations ($T \geq 10°C$) [f]		
BOD_5 (mg/L)	< 3	30 – 40
COD (mg/L)	< 20	80 – 100
TSS (mg/L)	< 5	10 – 20
NH_4-N (mg/L)	< 1	10 – 20

[a] Organic load: Q_i . COD effluent concentration from septic tank
[b] Required surface area: Organic load/Maximum organic loading rate
[c] Amount of a single load: Q_i/(24 hours/Loading interval)
[d] Surface area of intermittent loading tank: Q_i/(24 hours/Loading interval)
[e] Height difference in intermittent loading tank: Volume of a single dose/Surface area of dosing chamber
[f] Langergraber et al. (2007) for main layer of sand with gravel size 0.06 – 4 mm, and Langergraber *et al.* (2008) for main layer of coarse sand with grain size 2 – 3 mm, respectively.

4.5 CASE STUDY

The case study shows the oldest VF wetland implemented in Austria (Haberl *et al.*, 2003). The VF wetland was designed for 8 PE and treats wastewater of a farmhouse in Wolfern, Upper Austria. The farmhouse is too far away from the village to be connected to the sewer line. Therefore, it was selected within a pilot project as an example for many other farms in this area. The VF wetland was constructed in spring/summer of 1991 and has been operational since then. The effluent standards to be met are 25 mg BOD_5/L, 90 mg COD/L, 30 mg TSS/L, and 10 mg NH_4-N/L (at outlet wastewater temperatures > 12°C).

The system comprises a settling tank (volume 3 m^3) as primary treatment, followed by a storage tank for the intermittent loading (volume 2.7 m^3). The intermittent loading was with a pump controlled by an automatic valve that opened four times a day. Since 1998, the wastewater is loaded intermittently by means of a syphon. The main layer of the 42 m^2 VF bed has a depth of 80 cm and is filled with sand/gravel with grain size of 0.06 – 8 mm. The distribution of the wastewater on the surface is by PVC pipes with 8 mm holes. Below the holes, plates are situated to prevent erosion. The VF bed is planted with common reed (*Phragmites australis*). Figure 4.2 shows a schematic sketch of the system.

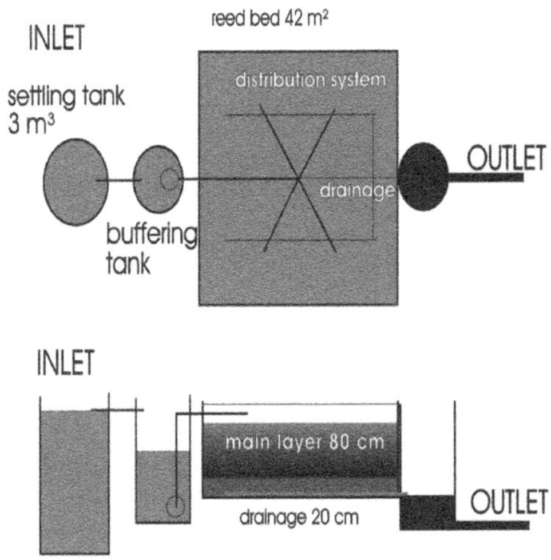

Figure 4.2 Schematic of the VF wetland for a farmhouse. Reprinted with permission from Haberl *et al.* (2003).

Table 4.3 shows performance data from the first six years of operation of the VF wetland. During this period, the mean HLR was 30 mm/d. The Austrian effluent standards (i.e. 25 mg BOD_5/L, 90 mg COD/L and 10 mg NH_4-N/L, and 95% and 85% for BOD_5 and COD, respectively) could be easily met.

Table 4.3 Performance from 1992 to 1997 of the VF wetland for a farmhouse - figures are yearly averages (Haberl *et al.*, 2003).

		BOD_5			COD			NH_4-N			PO_4-P		
Year	**HLR (mm/d)**	**In (mg/L)**	**Out (mg/L)**	**Removal (%)**	**In (mg/L)**	**Out (mg/L)**	**Removal (%)**	**In (mg/L)**	**Out (mg/L)**	**Removal (%)**	**In (mg/L)**	**Out (mg/L)**	**Removal (%)**
1992	24	143	11	92	378	54	86	48	8.6	82	10.8	3	72
1993	32	186	9	95	533	47	91	88	16.4	81	12.5	3.6	71
1994	27	139	3	98	366	36	90	71	1.3	98	11.5	6.1	47
1995	30	120	3	98	383	30	92	63	4.9	92	14	7	51
1996	37	157	<3	99	436	30	93	49	1.5	97	9.4	6.2	34
1997	30	278	<3	99.6	549	25	95	59	5.5	91	9.6	4.9	49
Average	**30**	**171**	**5**	**97**	**441**	**37**	**91**	**63**	**6.4**	**90**	**11.3**	**5.1**	**54**

The phosphorus elimination decreased from 72% to about 40 – 50% within the first years due to limited adsorption sites of the substrate. As reported by Laber et al. (1997), nitrified effluent was recycled into the settlement tank of the pre-treatment in 1995 (compare Section 6.3). An 80% recirculation rate increased the TN elimination to 72% (from originally 40% without recirculation). The VF bed is still in operation and performing well after more than 25 years.

5

French vertical flow wetlands

5.1 INTRODUCTION AND APPLICATION

In France, VF wetlands for treating raw wastewater have been introduced and successfully implemented. These systems treat sludge and wastewater in a single step. French VF wetlands are comprised of two stages, and each stage contains alternately operated cells. In the first stage, sludge treatment, partial removal of organic matter and nitrification takes place. In the second stage, final organic matter removal and nitrification occurs. The treated sludge from the first stage collects at a rate of approximately two to three cm per year when the system is operated at design load. The deposit layer must be removed once it reaches a depth of approximately 20 cm, which in practice, generally is every 10 to 15 years. The deposit layer may collect more slowly in systems that do not receive the full design load at the start of operation.

Over 4,000 systems have been constructed in France, with most systems serving populations of less than 1,000 PE. As of 2015, treatment wetlands represented more than 20% of all domestic wastewater treatment plants in France. Morvannou *et al* (2015) present a performance evaluation of 415 of these systems operating in France, with performance data from the first and second stages.

In the past decade, the French VF wetland design has also been implemented outside of France (in tropical overseas French territories, South America, as well as other countries within the European continent). The largest French wetland constructed is in Moldova and serves 20,000 PE (Masi *et al.*, 2017a). The application of wetland technology is generally not a question of feasibility but rather of economics and life cycle costs (including, but not limited to, the cost of raw materials, labour, construction, and operations and maintenance). In mainland France, treatment wetlands are generally economically favourable up to a size of 5,000 PE. For systems larger than 5,000 PE, conventional wastewater treatment technologies such as activated sludge (AS) plants, begin to be a better economical choice.

The first French treatment wetland guidance document was published in 2005 (Molle *et al.*, 2005). It includes guidance for what is today referred to as the classical two-stage French vertical flow wetland design (Figure 5.1). The first stage is fed with wastewater that has passed through a simple screen of 20 to 40 mm mesh. The first stage is divided in three parallel filters and the second stage in two filters.

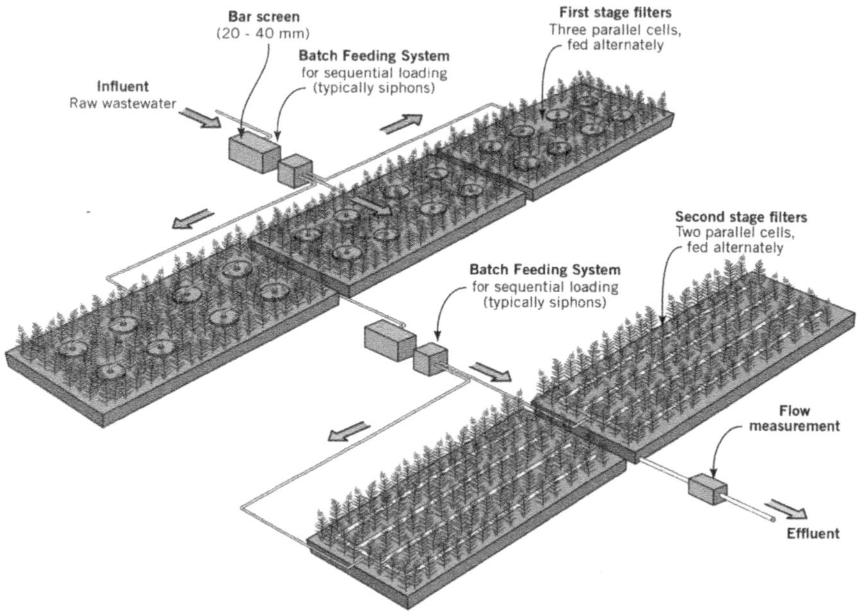

Figure 5.1 Schematic of the classical French VF design. Image courtesy of Epur Nature.

An important aspect of the system is its simplicity. Apart from the screen and the filters in the two stages, there are no other treatment units: no primary or septic tanks, no other biological treatment, no secondary sedimentation tanks or similar units and no sludge treatment units (since there are no other units that produce primary or excess sludge). A critical component of the French VF design is a well-established stand of *Phragmites*. As the deposit layer of organic matter develops, the reeds play a critical role. A small opening in the deposit layer is created as the plants stems are moved by the wind. This seemingly small detail is very important in maintaining the infiltration capacity of the filter, thus helping to maintain passive aeration of the filter. In other countries, other plants have been successfully used, but it is always important to analyse whether they will be able to undertake this function, and also resist periods without wastewater flow.

If the system is not expected to receive the full design flow at start-up, the wetland cells can be divided into smaller sections during the design process, so that portions of cells can be individually planted and loaded until the full design load is reached. This reduces the risk of poor plant establishment due to water stress.

Hydraulic considerations

As commented in Chapter 4 and emphasised in several places in this volume (mainly in Chapter 2), the unsaturated conditions of the filter medium in VF units imply that several hydraulic considerations (including the concept of retention time), modelling approaches and design specifications that are exclusive for saturated media (e.g. HF) are not applicable here. The reader is referred to these observations in the preceding text in order to avoid using inappropriate methods for the design of intermittently loaded VF wetlands.

The filters are dosed on an alternating basis, so that one filter is fed while the others are rested (Figure 5.2). These alternations are fundamental for proper operation of the French VF wetland. Alternating feeding helps control the growth of attached biomass on the surface of the filter media, helps maintain aerobic conditions within the filter bed itself, and aids in mineralisation of the organic matter that accumulates on the surface of the first stage filters. First stage filters are generally fed for 3.5 days and rested for 7 days. Second stage filters are generally fed for 3.5 days and rested for 3.5 days. The feeding pattern requires the system operator to visit the treatment plant twice per week, to switch the feeding, and ensure the system is functioning properly. The number of operator visits could be reduced with the installation of a Programmable Logic Controller (PLC) feeding system.

The alternating feeding pattern helps to:

- Ensure adequate oxygen transfer into the porous media.
- Stabilise the deposit layer on top of the filter beds in the resting stage.
- Implement resting phases that do not result in plant stress (e.g., lack of water)
- Maintain snow cover for maximum heat insulation during dry periods in cold climates. Dosing twice per week (as opposed to once per week) results in a shallower ponding depth of wastewater on the surface of the bed, which helps prevent the snow cover from melting (Prost-Boucle *et al.*, 2015).

In tropical climates, warmer temperatures allow for faster biological activity. As a result, the organic deposit does not require an entire week to be stabilised. This, combined with often less stringent effluent quality requirements results in the alternating feeding being kept at twice a week, but with only two filters on the first stage (Molle *et al.*, 2015). Filters are dosed in batches. The volume of a single batch must be between 2 and 5 cm in the filter in operation to ensure adequate distribution of water across the filter surface. The upper limit of 5 cm minimises the risk of preferential flow (short-circuiting). Figure 5.3 illustrates the operational sequence during one batch.

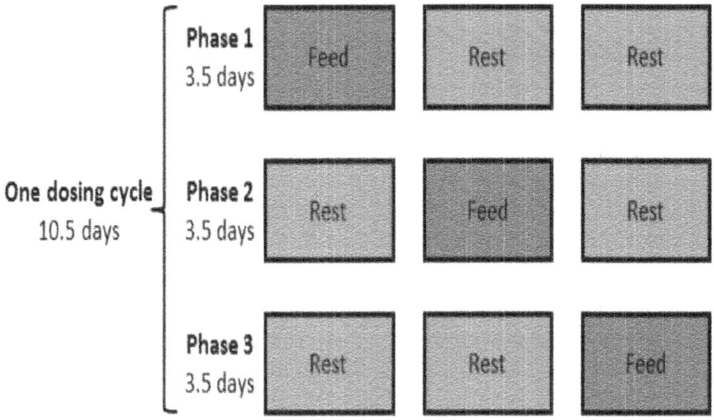

Figure 5.2 Operational scheme for the first stage of the French system, with three beds alternating the periods of feed and rest over a full cycle. In France, each phase is typically for 3.5 d and the full cycle is for 3x3.5 = 10.5 d (3.5 d feeding, 7.0 d resting).

To maintain aerobic conditions in the filter, passive oxygenation at the bottom of the filter is necessary. Drainage pipes (minimum diameter 100 mm) contain slots (length: ⅓ of pipe circumference, width: greater than 8 mm) for every 10 cm of drainage pipe length. The slots are positioned to allow air to enter via the top of the drainage pipes while treated wastewater is collected at the bottom of the drainage pipes.

Water distribution is different during each treatment stage (Figure 5.4). At the first stage, large pipes (> 110 mm diameter for small-scale systems, 160 – 200 mm diameter pipes for larger systems) are used to distribute the raw wastewater to one feeding point per 50 m^2. At the second stage, smaller pipes (> 110 mm diameter) with drilled holes (> 8 mm diameter) are used. The pipes for the second stage filters are laid directly on the filter surface. For both first and second stage filters, care must be taken to avoid scouring of the uppermost filtration layer. On first stage filters, a minimum flow of 0.5 $m^3/h{\cdot}m^2$ per batch is necessary to correctly distribute the water, while on second stage filters, the residual water pressure (or squirt height) at the outermost orifices must be greater than or equal to 30 cm.

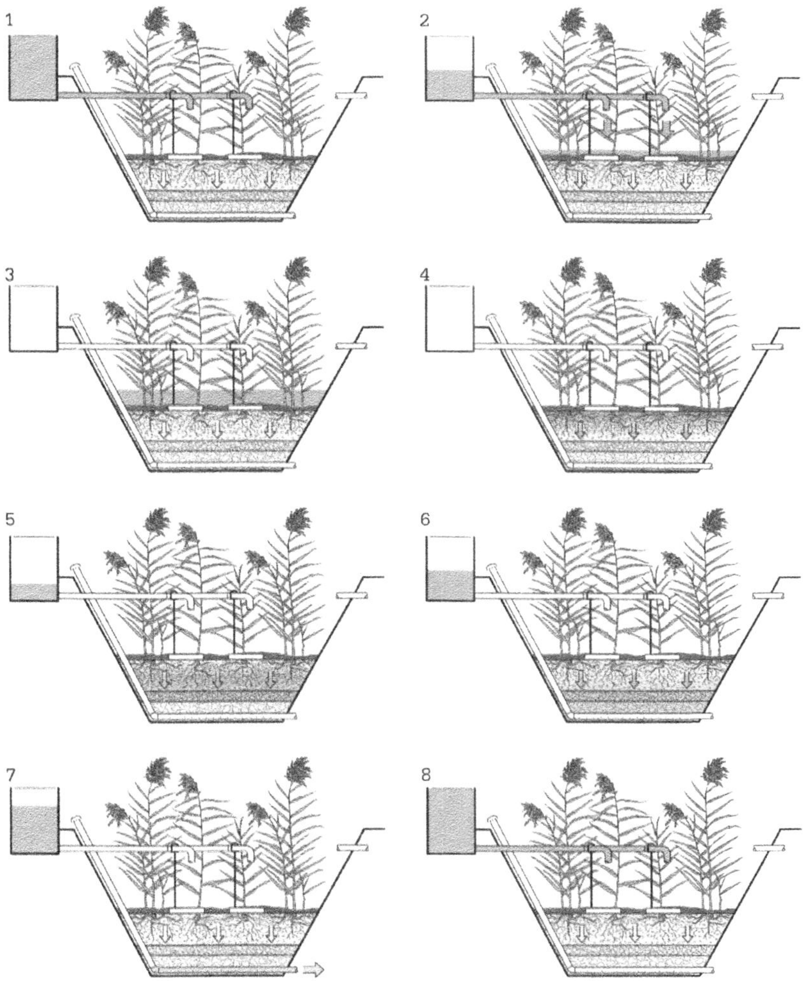

Figure 5.3 Operational sequence for one cell of a French VF wetland during a batch load to the filter in operation. Sequence moves by rows from top to bottom.

Figure 5.4 Distribution of wastewater French VF wetland; top: first stage; bottom: second stage. Photos courtesy of Pascal Molle.

5.2 DESIGN AND WATER QUALITY TARGETS

The design is based on maximum pollutant and hydraulic loads, which are expressed per m^2 of filter in operation per day (Table 5.1). For typical situations in France (in a temperate climate with separate municipal and storm sewers), this design leads to an area requirement of 0.4 m^2/PE for each filter cell, or in other words 1.2 m^2/PE for the first stage (with three cells) and 0.8 m^2/PE for the second stage (with two cells). The daily load of a PE is defined as 150 L wastewater, 157 g COD, 60 g BOD_5, 72 g TSS, 11.5 g NH_4-N, and 2.1 g TP (Mercoiret, 2010).

Table 5.1 Maximum design loads for classical French VF wetland design under dry weather conditions. Values given are per square meter of bed in operation.

Treatment Stage	HLR ($m^3/m^2 \cdot d$)	COD ($g/m^2 \cdot d$)	BOD_5 ($g/m^2 \cdot d$)	TSS ($g/m^2 \cdot d$)	TKN ($g/m^2 \cdot d$)
First stage	0.37	350	150	150	30
Removal [a]		$0.80 \times M_i$	$0.90 \times M_i$	$0.90 \times M_i$	$1.1128 \times M_i^{0.8126}$
Second stage	0.37	70	20	30	15
Removal [b]		$0.75 \times M_i$	$0.80 \times M_i$	$0.80 \times M_i$	$1.194 \times M_i^{0.8622}$

[a] All correlations from Molle et al (2005), except TKN from Molle et al (2008), M_i stands for mass load in $g/m^2 \cdot d$

[b] All correlations from Molle et al (2005)

The specified loading rates are for one filter in operation. When comparing loading rates applied in other treatment systems, these values should be divided by three in the first stage and by two in the second stage to take into account the entire treatment area.

The systems treat screened domestic wastewater. The first stage filters are effective for removal of organic matter and TSS. The second stage filters have a polishing effect for COD, BOD_5 and TSS. Designing and operating the wetlands within the design envelope specified here results in systems that can guarantee final effluent concentrations of 90 mg COD/L, 20 mg BOD_5/L, 15 mg TSS/L, and 15 mg TKN/L. While organics and suspended solids follow a linear removal trend, TKN removal efficiencies are more complex. The equations provided allow calculation of estimated removal efficiencies when operating the filters within the design envelope (Molle *et al.* 2005, Molle *et al.* 2008).

Filter media specifications

The filters in the first stage and the second stage use different media and have different dimensions to provide the conditions conducive of treatment under the design loads (Figure 5.5). To ensure aerobic conditions in the first stage filter, the main layer is composed of 2 – 6 mm gravel. A smaller grain size would lead to clogging and a coarser grain size would hinder the formation of the organic deposit layer. As in VF wetlands, below the main filter layer, a transition or intermediate layer (5 – 15 mm gravel) prevents finer particles from being washed into the drainage layer (thus reducing the effective porosity of the drainage layer). Treated water is collected in drainage pipes in the drainage layer, which consists of coarse gravel (20 – 60 mm) at the bottom of the bed. The filters are isolated from the surrounding land by a combination of a plastic liner and a geotextile membrane.

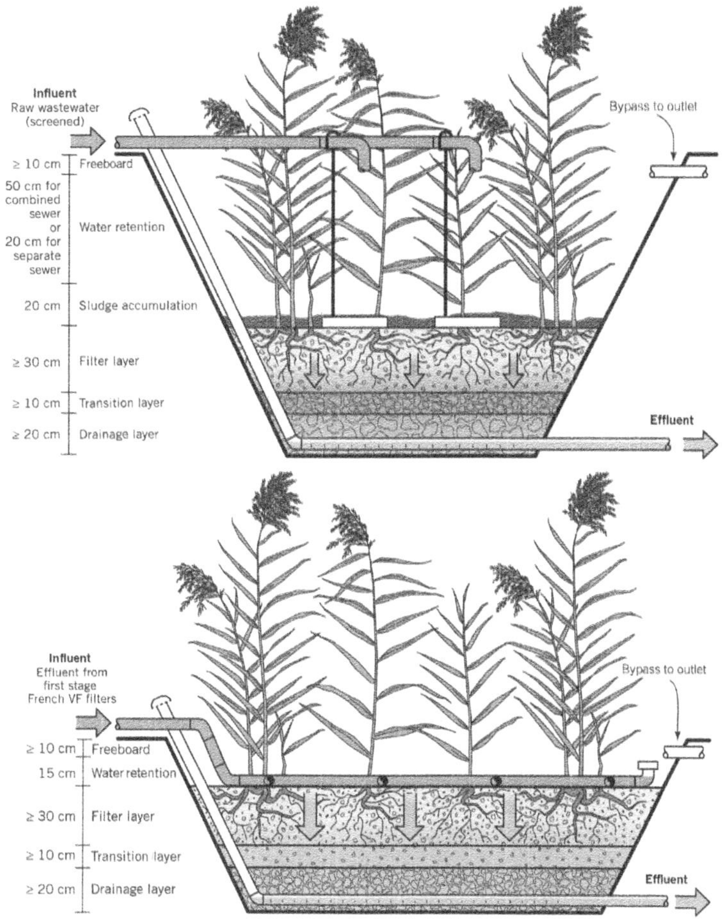

Figure 5.5 Profile of French VF cells; top: first stage; bottom: second stage.

Sand is the main filter media for the second stage beds (Table 5.2) defined as a compromise between good performance removal while not clogging (Liénard *et al.*, 2001). Washed, crushed sand can also be used but has been found to lead to lower treatment performance (Torrens *et al.*, 2009). Consequently, a deeper layer of sand must be used if the sand specifications in Table 5.2 cannot be met. For both the transition and drainage layer, the Terzaghi rule is used ($D_{15}/d_{85} \leq 4$). A permeability criterion is also added: $D_{15}/d_{15} \geq 4$. This rule is used to ensure that the interface between the sand layer and transition layer does not produce a decrease in permeability by reducing the local porosity.

Table 5.2 Filter media specifications for a French VF wetland design. Adapted from Molle *et al.* (2005).

	First stage		**Second stage**	
	Depth	**Material**	**Depth**	**Material**
Freeboard	> 30 cm		> 20 cm	
Main layer	30 to 80 cm	2 – 6 mm gravel	30 to 80 cm	sand $0.25 < d_{10} < 0.4$ mm *and* $d_{60}/d_{10} < 5$ *and* less than 3% fine particles
Transition layer	10 to 20 cm	5 – 15 mm gravel	10 to 20 cm	3 – 12 mm gravel
Drainage layer	20 to 30 cm	20 – 60 mm gravel	20 to 30 cm	20 – 60 mm gravel

Construction of the cells is typically with a surface length-to-width ratio of one, with an embankment slope of 1:1. This means, unlike HF systems, the total volume of media that will be required is less than what would be calculated from a square cell.

The depth of the main layer will impact the performance reliability. The bulk of carbon and ammonium removal in French VF wetlands occurs within the uppermost 10 – 40 cm of an unsaturated filter (Millot *et al.*, 2016). When stringent effluent concentrations must be met, the depth of the main layer can be increased (> 60 cm for COD removal and > 80 cm when full nitrification is required).

5.3 OPERATION AND MAINTENANCE

The O&M of French VF wetlands is comprised of different phases in which specific tasks must be performed. The phases include the commissioning period, the routine operation period and periods during which the organic matter is removed. A French guideline of O&M is available on the EPNAC website (www.epnac.irstea.fr).

Commissioning period

Good reed growth on the first stage is fundamental to maintain water infiltration capacity and passive aeration of the filter. During the first year, in addition to routine maintenance tasks (discussed in the next section), attention must be given to avoid excessive weed growth in the filters. It is a manual and fastidious task. It is possible to saturate the filter for one or two weeks during the first growing season to kill the weeds and favour reed establishment and growth. In this case, it is better to not saturate the first and second stages at the same time so as not to hinder the nitrification process.

Troubleshooting might be necessary during commissioning periods. The main issues that occur during commissioning include:

- When the system starts with a very low hydraulic load, water infiltrates near the loading points and the reeds can suffer from water stress. This does not cause a problem for removal performance, but weed maintenance can be an arduous task.
- When a system is started at its nominal design load, the organic matter deposit forms quickly. This is because the reeds are too small to aid in water infiltration and deposit mineralisation. The deposited organic matter therefore dries quickly, without mineralisation, and can contribute to excess ponding. This problem ends once the plant stand becomes established.
- When the deposit layer is not yet developed and the treatment plant receives storm events, water can infiltrate quickly into the first stage and ponding (event surface clogging) can appear on the second stage. This phenomenon ends once the deposit layer appears on the first stage. A sludge or compost layer can be applied on the first stage to accelerate the process.

Routine maintenance

The operator should visit the site twice a week to check the functioning of the system and to perform specific short tasks. The coarse screening (prior to the first stage cells) must be cleaned regularly, and the batch feeding systems have to be checked for proper operation. Filter alternation must be done every 3.5 days to maintain good oxygen content in the filter. Rest periods that are too long are unfavourable for the microbial community in the wetland. Other maintenance tasks can be done less frequently, such as controlling the weeds (once a month) or checking the organic deposit height and harvesting the reeds (once a year). In tropical climates, the maintenance frequency of the plant stand can be higher due to the warmer climate and thatch accumulation.

Removal of accumulated organic matter

The deposit layer increases approximately 2 to 3 cm/yr on the first stage filters, when fed at the design load in a temperate climate (in tropical climates, even if using only two filters in parallel, the accumulation is slower). This deposit layer is a hydraulically limiting step and must be removed once reaching 20 cm (generally 10 – 15 years). If the deposit layer is not removed when it reaches a depth of 20 cm, ponding will occur and oxygen transfer to the subsurface will be hindered. Due to the mineralisation over many years, the deposit layer has a dry matter content greater than 25% and an organic matter content of approximately 40%. This organic matter is removed by a mechanical machinery and can be spread onto fields as an organic matter and phosphorous source, depending on local regulations. Contrary to sludge treatment wetlands (see Section 8.3), there is no need to apply a specific rest period before sludge removal. The French VF wetlands can be put back in operation immediately after sludge removal has been completed.

Load variations

Load variations (organic or hydraulic) can affect performance of the filters. Organic overloads increase oxygen consumption and saturate ammonium adsorption sites faster. This should be avoided, because ammonium adsorption onto organic matter is an important key parameter in nitrification (Morvannou *et al.*, 2014). Nevertheless, higher organic loads can be applied in summer because of the higher biological kinetic rates. This means the system does not need to be oversized in order to account for variations in the organic load during the summer season, when population may increase in tourist areas (Boutin and Prost-Boucle, 2015).

Hydraulic overloads can induce longer ponding periods that also may affect oxygen transfer into the deposit layer and main filter layer. At the same time, hydraulic overloads decrease the water retention time within the filter (Molle *et al.*, 2006). With proper design, storm events can be treated in the filter while maintaining acceptable outlet concentrations (Arias, 2013).

5.4 DESIGN EXAMPLE

Design a French VF wetland for a small community (100 PE) in a temperate climate to deliver a final effluent quality of 90 mg COD/L, 20 mg BOD_5/L, 15 mg TSS/L, and 15 mg TKN/L.

Input data:

- Average wastewater generation per PE:150 L/d.
- Average daily mass loadings per PE: 150 g COD, 60 g BOD, 70 g TSS, and 15 g TKN.
- Screened sewage will be applied on the wetland units.
- Target effluent: 90 mg COD/L, 20 mg BOD_5/L, 15 mg TSS/L, and 15 mg TKN/L.

Step 1. Define influent flow and pollutant concentrations

$$\text{Inflow, } Q_i = 100 \text{ PE} \times 150 \frac{\text{L}}{\text{PE} \cdot \text{d}} \times \frac{1 \text{ m}^3}{1000 \text{ L}} = 15 \frac{\text{m}^3}{\text{d}}$$

$$\text{BOD Mass Load In}, M_i = 100 \text{ PE} \times 60 \frac{\text{g BOD}}{\text{PE}\cdot\text{d}} = 6{,}000 \frac{\text{g BOD}}{\text{d}}$$

$$\text{BOD Concentration In}, C_i = 6{,}000 \frac{\text{g BOD}}{\text{d}} \div 15 \frac{\text{m}^3}{\text{d}} = 400 \frac{\text{g BOD}}{\text{m}^3} = 400 \frac{\text{mg BOD}}{\text{L}}$$

Doing similar calculations for all pollutants, the influent loads and concentrations are found for the design of the wetland system (Table 5.3).

Table 5.3 Design example influent characteristics.

Variable	BOD	COD	TSS	TKN
Average daily load per PE (g/d)	60	150	70	15
Influent load (g/d)	6,000	15,000	7,000	1,500
Influent concentration (mg/L)	400	1,000	467	100

Design of the first stage of the system

Step 2. Calculate required areas based on recommended hydraulic and mass surface loading rates

The recommendations on loading rates for the first stage of the French VF system (Table 5.1) are used to provide initial sizing of the filter units, as follows:

Required surface area of one filter based on a HLR of $q = 0.37$ m^3/m^2·d:

$$A_1 = \frac{Q_i}{q} = \frac{15 \text{ m}^3/\text{d}}{0.37 \text{ m}^3/(\text{m}^2.\text{d})} = 41 \text{ m}^2$$

Required surface area of one filter based on a surface mass loading rate of BOD of $M_i = 150$ g/m^2·d:

$$A_1 = \frac{M_i}{M_i} = \frac{6{,}000 \text{ g/d}}{150 \text{ g}/(\text{m}^2.\text{d})} = 40 \text{ m}^2$$

Doing similar calculations for all parameters considered and adopting the surface loading rates shown in Table 5.1 for the first stage of the French VF system leads to five different required surface areas, namely, 41 m^2, 40 m^2, 43 m^2, 47 m^2, and 50 m^2, based on hydraulic, BOD, COD, TSS and TKN loads, respectively. The largest one (50 m^2) will be chosen to enable compliance with all the design criteria and produced required effluent concentrations.

For square cells, a length-to-width ratio of $l/w = 1$ will lead to:

$$A = l \times w = w^2$$

$$w = \sqrt{A} = \sqrt{50} = 7.1 \text{ m}$$

Adopting values of length $l = 7.5$ m and width $w = 7.5$ m will lead to an area of 56.2 m^2 per unit.

Considering that the first stage of the French VF system is comprised of three units in parallel (n=3, with one working and two resting), the total area required for the first stage is:

$$A_T = n \times A_1 = 3 \times 56.2 \text{ m}^2 = 169 \text{ m}^2$$

The total area of 169 m^2, for a population of 100 PE, corresponds to a net per capita land requirement of 169/100 = 1.69 m^2/PE. This is higher than the usual French VF wetland practice (1.2 m^2/PE; 0.4 m^2/PE for each filter), as it is limited by TKN rather than BOD removal requirements.

Step 3. Specify the characteristics of the filter layers

The filter layers will follow the French specifications, with intermediate values from the ranges of height for each layer (Table 5.2). This results in:

- Freeboard: h = 0.50 m.
- Main layer: h = 0.50 m (2 – 6 mm gravel).
- Transition layer: h = 0.15 m (5 – 15 mm gravel).
- Drainage layer: h = 0.25 m (20 – 60 mm gravel).
- Total filter height: h_T = 0.50 + 0.50 + 0.15 + 0.25 = 1.40 m.

With the total filter area of 169 m^2, the volumes of support material to be used are calculated based on a 1:1 sloped cell, using the average top length and width for each layer. This results in:

- Main layer: V=12.3 m^3. For three cells, V=37 m^3.
- Transition layer: V=3.0 m^3. For three cells, V=9 m^3.
- Drainage layer: V=4.4m^3. For three cells, V=13 m^3.

Step 4. Determine the characteristics of each batch

The water level that will cover the filter during each batch corresponds to the volume of each batch divided by the surface area of the filter in operation. The French VF wetland recommendations are for a liquid layer between 2 and 5 cm (for the purposes of this example, 3 cm will be considered for the batch load).

First, the volume per batch is calculated as follows:

$$3 \text{ cm} \times \frac{1 \text{ m}}{100 \text{ cm}} \text{ (batch load in m)} = \frac{\text{Volume per batch (m}^3)}{\text{Area of filter bed in operation (m}^2)}$$

$$\text{Volume per batch (m}^3) = 56.2 \text{ m}^2 \times 0.03 \ \frac{\text{m}}{\text{batch}} = 1.7 \ \frac{\text{m}^3}{\text{batch}}$$

Then the number of required batches per day is calculated as follows:

$$\text{Number of batches per day} = \frac{15\ \text{m}^3/\text{d}}{1.7\ \text{m}^3/\text{batch}} = 9\ \frac{\text{batches}}{\text{d}}$$

To calculate the wastewater flow during the batch, the minimum instantaneous HLR of 0.5 $m^3/m^2·h$ is used:

$$\text{HLR batch} = \frac{\text{Batch flow}}{\text{Area of filter bed in operation}}$$

$$\text{Batch flow} = 56.2\ \text{m}^2 \times 0.5 \frac{\text{m}^3}{\text{m}^2 \cdot \text{h}} = 28.1 \frac{\text{m}^3}{\text{h}}$$

A rounded-up value of 30 m^3/h will be used for this system. This results in an instantaneous HLR of 0.53 $m^3/m^2·h$, which gives an extra safety factor and is an easier working number.

The flow of 30 m^3/h is equivalent to (30 m^3/h) / (60 min/h) = 0.5 m^3/min. Since the volume of each batch is 1.7 m^3 (see previous calculation), the duration of each pulse will be (1.7 m^3) / (0.5 m^3/min) = 3.4 minutes. This is within the typical range of values for a pulse feeding during a batch.

Design of the second stage of the system

Step 5. Calculate the influent characteristics

The effluent quality from the first stage is needed to use as influent to the second stage filters. The expected quality is calculated using the removal rates from Table 5.1, as follows:

$$\text{BOD removed in first stage} = 0.9 \times MLR_{BOD} = 0.9 \times 120 \frac{\text{g BOD}}{\text{m}^2 \cdot \text{d}} = 108 \frac{\text{g BOD}}{\text{m}^2 \cdot \text{d}}$$

$$\text{Effluent BOD areal daily mass from first stage} = 120 \frac{\text{g BOD}}{\text{m}^2 \cdot \text{d}} - 108 \frac{\text{g BOD}}{\text{m}^2 \cdot \text{d}} = 12 \frac{\text{g BOD}}{\text{m}^2 \cdot \text{d}}$$

$$\text{Effluent BOD mass flow from first stage} = 12 \frac{\text{g BOD}}{\text{m}^2 \cdot \text{d}} \times 50\ \text{m}^2 = 600 \frac{\text{g BOD}}{\text{d}}$$

$$\text{Effluent BOD concentration from first stage} = 600\ \frac{\text{g BOD}}{\text{d}} \div 15\ \frac{\text{m}^3}{\text{d}} = 40\ \frac{\text{g BOD}}{\text{m}^3} = 40\ \frac{\text{mg BOD}}{\text{L}}$$

The calculations are repeated for each parameter. The resulting effluent quality from the first stage is 40 mg/L, 200 mg/L, 47 mg/L, and 41 mg/L for BOD, COD, TSS, and TKN, respectively.

Step 6. Dimension the filters and calculate final effluent concentrations

The typical second stage of the French VF system has two filters in parallel, alternating their operation every 3.5 days. This will be also adopted here (n = 2). The dimensions of each filter are based on the maximum loading rates for the second stage following the same procedure as the first stage (Table 5.4). This includes the iterations and final dimensioning based on limiting area, i.e., 43 m^2 resulting in cells of 49 m^2. The final effluent concentrations are calculated based on the removal efficiencies from Table 5.1.

Table 5.4 Second stage filter dimensioning and performance

Variable	Flow	BOD	COD	TSS	TKN
Influent flow (m^3/d) and concentrations to the 2nd stage (mg/L)	15	40	200	47	41
Maximum recommended surface loading rates according to French specifications	0.37 $m^3/m^2 \cdot d$	20 $g/m^2 \cdot d$	70 $g/m^2 \cdot d$	30 $g/m^2 \cdot d$	15 $g/m^2 \cdot d$
Required area for filter in operation (m^2)	41	30	43	23	41
Actual areal loading for filter in operation	0.31 $m^3/m^2 \cdot d$	12.2 $g/m^2 \cdot d$	61.2 $g/m^2 \cdot d$	14.3 $g/m^2 \cdot d$	12.6 $g/m^2 \cdot d$
Final effluent flow (m^3/d) and concentrations (mg/L)	15	8	50	9	7

The effluent quality produced by the design adopted here meets the discharge requirements and is better than the average performance of a typical French VF treatment plant. This is because sizing was rounded upward from the most restrictive criteria, TKN. As with other treatment technologies, the designer can choose the degree of risk and safety in the design parameters chosen. It is strongly recommended to avoid a design that delivers effluent quality too close to the guaranteed limits.

Step 7. Specify the characteristics of the filter layers

The filter layers will follow the French specifications, with intermediate values from the ranges of height for each layer (Table 5.2). The height of the layers will be the same as those from the first stage, with the exception of the freeboard, which will be smaller (0.30 m), resulting in:

- Freeboard: $h = 0.30$ m.
- Main layer: $h = 0.50$ m (sand 0.25mm $< d_{10} <$ 0.4mm and $d_{60}/d_{10} < 5$).
- Transition layer: $h = 0.15$ m (3 to 12 mm gravel).
- Drainage layer: $h = 0.25$ m (20 – 60 mm gravel).
- Total filter height: $h_T = 0.30 + 0.50 + 0.15 + 0.25 = 1.20$ m.

With the total filter area of 98 m^2, the volumes of support material to be used are calculated based on a 1:1 sloped cell, using the average top length and width for each layer. This results in:

- Main layer: V=10.6 m^3. For two cells, V=21.1 m^3.
- Transition layer: V=2.6 m^3. For two cells, V=5.1 m^3.
- Drainage layer: V=3.7m^3. For two cells, V=7.4 m^3.

Step 8. Make schematic drawings of the system

The schematic arrangement, seen in top view, is shown in Figure 5.6, whilst the filter layers of both stages and their thickness are summarised in Table 5.5.

Table 5.5 Filter layers and layer thickness (in cm) in the first and second stages of the French VF wetlands of this example.

Variable	First stage	Second stage
Freeboard	50	30
Main layer	50	50
Transition layer	15	15
Drainage layer	25	25

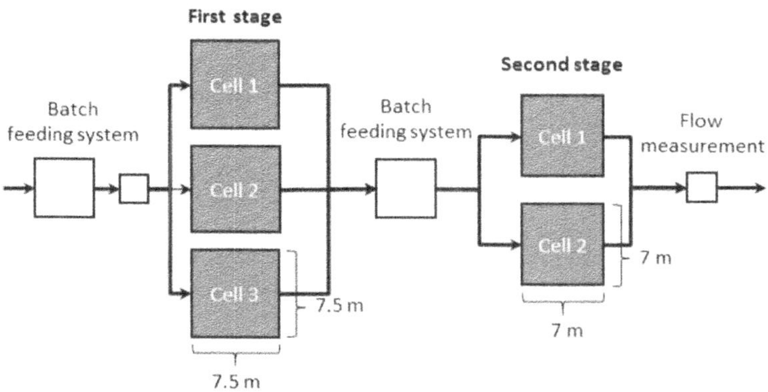

Figure 5.6 Schematic arrangement of the filter units in the two stages of the French VF wetlands of this example.

5.5 CASE STUDY

As an example, we present the treatment plant of Roussillon, France. The town of Roussillon had to replace an old trickling filter system that was used to treat the wastewater. The town of Roussillon has a very large tourist population in the summer, and is in a very sensitive environment. No electricity was available on site. The design capacity of the treatment wetland system is 1,250 PE. The effluent standards to be met were 25 mg BOD_5/L, 125 mg COD/L, 35 mg TSS/L. Nitrification requirements were not stringent (50%). However, the design was made to achieve effluent concentrations below 15 mg/L of TKN, in case more stringent effluent requirements would be imposed in the future.

The system is comprised of a screen (2 cm), followed by a storage tank for intermittent loading (volume 5.4 m^3). The intermittent loading works by gravity, using a siphon. The first stage of wetland cells is comprised of three filters in parallel for a total surface of 920 m^2. The main layer of the filter is 60 cm deep and contains gravel with a nominal grain size of 2 – 6 mm. The distribution of the wastewater on the surface of the wetland is achieved using stainless steel pipes of 160 mm diameter (8 feeding points per filter). Below the feeding points, concrete slabs are situated to prevent surface erosion. There is a second storage tank (volume 6 m^3) and siphon downstream of the first stage which provides intermittent loading to the second stage wetland cells. The second stage is comprised of two filters in parallel for a total system surface of 500 m^2. The second stage is smaller than usual design because of the high variation in load. The main layer of the second stage has a depth of 60 cm and is filled with sand (d_{10} = 0.4 mm; d_{60} = 1 mm). The distribution of the wastewater on the surface is achieved via PVC pipes with 8 mm holes. Some gravel is placed under the distribution orifices to prevent erosion. All the filters are planted with common reed.

The wetlands blend completely into the valley (Figure 5.7), which is visited by many tourists. Since the system was put into operation in 1998, the plant has been treating an average pollution load of about 950 PE in the high summer season and of 710 PE in the low winter season.

Table 5.6 shows performance data from the first 15 years of operation of the treatment wetland. During this period, the mean HLR was 20 cm/d on the filter in operation, and up to 50 cm/d in summer. The outlet requirements are met even if the organic load sometimes exceeded the design load in summer (up to 400 g COD/m^2·d at the end of August). The treatment plant is still in operation and performing well after its first 18 years.

Figure 5.7 Aerial view of the Roussillon French VF wetland. Photo courtesy of Vincent Leboeuf, SYNTEA group.

Table 5.6 Performance of the Rousillon French VF wetland from 1998 to 2012 (average of 24 h flow composite samples from the IRSTEA database). Hydraulic load shown as applied only to the first stage filter in operation.

	Hydraulic	BOD_5			COD			TSS			TKN		
	Load	In	Out	Removal	In	Out	Removal	In	Out	Removal	In	Out	Removal
	(cm/d)	(mg/L)	(mg/L)	(%)	(mg/L)	(mg/L)	(%)	(mg/L)	(mg/L)	(%)	(mg/L)	(mg/L)	(%)
Normal season	18.3	392	5	98	809	34	95	403	4	99	85.7	4.3	95
July/August	45.6	355	5	98	710	40	94	368	6	98	69.6	6.4	90

The sludge deposit layer was removed in 2011, after 13 years of operation (Figure 5.8). The sludge accumulation rate for this system was approximately 2.3 cm/yr. The deposit had a dry matter content of 45% and was well stabilised (organic matter content was approximately 40% of the dry matter content). Metals accumulation into the deposit layer was not problematic when compared to the French standard for direct agricultural application.

Figure 5.8 Sludge removal from the first stage French VF wetland in Roussillon, France after 13 years of operation (Troesch and Esser, 2012).

6

Intensified and modified wetlands

6.1 INTRODUCTION AND APPLICATION

The three treatment wetland types (HF wetlands, VF wetlands and French VF wetlands), as described in the previous chapters, are passive treatment systems that can be operated without external energy supply (with exception of pumps that might be required for loading). Over the last decade, new wetland designs and/or operational strategies have been developed in order to comply with higher water quality standards for phosphorus and nitrogen removal and to reduce surface area requirements. These new strategies have led to a group of wetland technologies that are collectively referred to as *intensified treatment wetlands*.

Intensified wetlands generally fall into one of two categories: use of an external energy source to increase the aerobic capacity of the system (such as the reciprocating operational strategy, effluent recirculation or forced aeration) or specific filter media (mainly aimed at the improvement of phosphorus removal). This chapter addresses the most common intensified operational and design strategies for the enhancement of nutrient (phosphorus and nitrogen) and organic matter removal.

6.2 REACTIVE MEDIA

Reactive media can be used in subsurface flow wetlands instead of sand and/or gravel to either enhance phosphorus removal or provide temporary sequestration of ammonium. Phosphorus removal in HF and VF wetlands for secondary treatment of domestic wastewater is mainly based on sorption, precipitation and crystallisation onto materials rich in Ca, Fe and Al. Reactive media for enhanced phosphorus removal include natural products, industrial by-products, and manufactured products. Table 6.1 shows some of the reactive media used for phosphorus retention as reported by Vohla *et al.* (2011). A main disadvantage of using reactive media is that most increase the pH of the effluent significantly as a by-product of the phosphorous removal (Table 6.1). Many commercially available products to enhance phosphorus removal have been used in treatment wetlands, however locally available materials are recommended over other materials whenever possible.

In general, calcium-based materials are widely used because calcium ions can form stable and insoluble products with phosphate. At lower phosphorus concentrations, adsorption is the dominant process for phosphorus removal, whereas at high phosphorus concentrations, precipitation takes place. Precipitates reduce the pore volume for the filter and over a long time can lead to reduced hydraulic performance and hydraulic failure due to clogging.

Table 6.1 Examples of reactive filter media used for phosphorus retention in treatment wetlands (extracted from Vohla *et al.*, 2011). Sand is shown as comparison as non-reactive filter material.

Media	**Phosphorus retention g P/kg media**	**Effluent pH**	**Comment**
Natural media			
Sand	0.1	7.4	HF wetland, full-scale
Apatite	14	7.0	550-day column experiment
Oyster shells (heated)	120	8.3	Batch test
Oyster shells (burnt)	830	12.6	Column test
Shell sand	335	8.6	HF wetland, 1 year
Industrial by-products			
Oil shale ash	25-60	> 12	Batch and pilot-scale experience
Blast furnace slags	1-2	> 10	Batch to full-scale experience
Fly ash	100-400	9 – 11	Batch to full-scale experience
Manufactured products			
Filtralite P ®	0.5-5	> 10	Batch to full-scale experience
Ferrosorp ®	up to 14	7 – 8	Batch to full-scale experience
LECA ®	up to 12	10 – 12	Batch to full-scale experience
Phosclean ®	up to 14	8.5 – 11	

Phosphorus removal with reactive media in treatment wetlands is a time-limited process mainly due to limited adsorption sites and partly also due to reduced hydraulic functioning caused by precipitates. After the finite capacity of the filter material is reached causing the effluent phosphorous concentrations to rise toward the effluent limit, the filter material must be changed. Instead of changing the whole filter material of the treatment wetland, it is advised to add unplanted phosphorus filters as a pre- or post-treatment step. However, smaller post-treatment filters will reach phosphorus saturation more quickly and, thus, need more frequent exchange of the filter material (Jenssen *et al.*, 2010). As phosphorus is a limited resource, filter materials from which phosphorus can be reused as fertiliser in agriculture are preferable. The effects of high phosphorus binding capacity for high removal capacity and high plant availability of phosphorus as fertiliser might be contradictory and is still a matter of research (Jenssen *et al.*, 2010).

Besides enhancing phosphorus removal, reactive media are also used for increasing the retention of ammonia. Zeolites have been applied in hydraulically higher loaded treatment wetlands to adsorb ammonia (and thus increasing its removal) during the loading phase of the system. The adsorbed ammonia is then desorbed and nitrified in the following phase without loading. Zeolites have been successfully used in VF wetlands (Dal Santo *et al.*, 2010; Canga *et al.*, 2011; Stefanakis and Tsihrintzis, 2012b) and French VF wetlands (Paing *et al.*, 2015).

6.3 RECIRCULATION

Recirculation involves returning and mixing a portion of the wetland effluent with the influent of the treatment plant. Effluent recirculation has been proposed as an operational modification to improve organic matter and nitrogen removal especially in highly aerobic VF systems. Removal of TN is enhanced because effluent with appreciable nitrate but limited organic matter is mixed with influent low in nitrate but high in organic carbon, allowing denitrification to take place. The principles of recirculation applied to treatment wetlands are similar to pre-denitrification schemes as applied in conventional activated sludge plants (Chapter 7 of Volume 5; von Sperling, 2007d).

Though most common with VF systems, recirculation has been applied to virtually all treatment wetland types. As early as the 1990's, experiments with recirculation on a VF wetland showed improved efficiency and resulted in recirculation being incorporated into some design guidelines (Laber *et al.*, 1997; Brix and Johansen, 2004; Brix and Arias, 2005). Recirculation ratios range generally from 50% to 200% of the influent flow. Higher ratios return more nitrate for additional denitrification but simultaneously increase the hydraulic loading and therefore decrease the HRT of the first pass influent, thus the proper recirculation ratio is specific to the hydraulic and nutrient loading rates of the system.

Table 6.2 shows examples of studies where recirculation was applied. For HF wetlands, the increased hydraulic load due to recirculation was not beneficial, and the removal efficiencies and removal rates decreased. For VF wetlands, the TN removal efficiency could be increased with higher recirculation rates, however, COD and NH_4-N removal efficiency decreased (Laber *et al.*, 1997). Foladori *et al.* (2014) showed that with recirculation, a VF wetland can be operated with higher organic and hydraulic loads and high removal efficiencies and rates can be achieved. Recycling applied to French VF wetlands has the main aim of reaching the legal requirements with only one stage and thus reducing the surface area requirement. Results showed that nitrification efficiency could be improved while enhancing organic matter removal. However, at high loads nitrification and organic matter removal are reduced again.

Table 6.2 Examples of recirculation applied to various treatment wetland types.

Type	Recirculation rate	Removal efficiency (%)			Removal rate ($g/m^2 \cdot d$)			Reference
		COD	NH_4-N	TN	COD	NH_4-N	TN	
HF	0%	88	80	81	9.0	0.5	0.9	Stefanakis and Tsihrintzis (2009)
	50%	86	41	65	6.9	0.2	0.6	
VF	0%	92	99	15	10.6	1.9	0.3	Laber et al. (1997)
	50%	91	99	39	10.5	1.9	0.9	
	80% [a]	88	97	64	10.1	1.8	1.4	
	0%	78	75	27	25.9	3.0	1.4	Foladori et al. (2014)
	100% [b]	86	87	39	58.5	5.2	3.2	
French VF (1st stage)	50%	90	58	56	175.9	6.0	9.7	Prost-Boucle and Molle (2012)
	100%	92	78	61	115.9	6.9	8.6	
	100% (high) [c]	80	41	23	140.8	5.0	4.1	

[a] Recirculation in primary treatment unit, with 20 cm water saturation in the main layer.
[b] Recirculation on top of the filter, with 20 cm water saturation in the bottom.
[c] 1.9-fold higher HLR and 1.4-fold higher organic loading rate, respectively.

6.4 PARTIAL SATURATION

Another way to increase TN removal is partial saturation of VF wetlands and French VF wetlands. Partial saturation means that the upper layer of a VF cell is unsaturated and the bottom layer is impounded. The unsaturated part of the system remains under aerobic conditions, which allows for good nitrification. The saturated part of the VF bed allows for denitrification if anoxic conditions prevail and enough organic matter reaches the saturated part of the treatment bed. Table 6.3 shows examples of partial saturation applied to a VF wetland and a single-stage French VF wetland, respectively. For a VF wetland with 50 cm main layer, TN removal can only be increased when coarse sand (e.g., media size 1 – 4 mm) is used for the main layer. If finer sand is used, organic matter is already degraded and not available anymore for denitrification in the 15 cm impounded drainage layer. For a single-stage French VF wetland, a 40 cm impounded layer increased removal of TN compared to a 20 cm drainage layer. However, when recirculation is added, there is no difference anymore regarding removal rates for different depths of the impounded layer.

6.5 RECIPROCATION

In treatment wetlands, sequential filling and draining of wastewater can be employed in order to increase subsurface oxygen availability, and thus removal of oxygen demanding compounds such as COD, BOD_5 and ammonium nitrogen. These treatment wetlands are commonly known as tidal flow, fill-and-drain, or reciprocating wetlands (Sun *et al.*, 1999; Behrends *et al.*, 2001; Austin, 2006; Wu *et al.*, 2011; Stefanakis *et al.*, 2014). Frequent water level fluctuation, or operation in fill-and-drain mode has been shown to increase treatment performance compared to beds with a static water level (Tanner *et al.*, 1999). Reciprocation refers to the alternate filling and draining of pairs of wetland cells, whereas tidal flow and/or fill-and-drain wetland cells can either be configured pairwise or in series (forward-flow only, one after another).

The rate of oxygen transfer into reciprocating treatment wetlands is related to the frequency of the water level fluctuation. During the drain cycle, air is drawn into the filter matrix and into the thin water film on the surface of the filter media (Green *et al.*, 1997). Diffusion of oxygen into the thin water film is rapid (on the order of seconds) (Behrends *et al.*, 2001). During the subsequent fill cycle, the thin water film on the gravel surface is surrounded by anaerobic or anoxic water, and reducing conditions prevail. The alternating oxic/anoxic sequence is repeated multiple times per day (between six and 24 cycles per day), which creates unique conditions that develop a microbial community that is diverse and robust. As a result, reciprocating wetlands are particularly well suited for removing pollutants from complex wastewaters, and have shown high removal rates, especially for TN (Table 6.4).

Table 6.3 Examples of partial saturation applied to various VF and French VF wetlands.

Type	Degree of saturation	Removal efficiency (%)			Removal rate (g/m²·d)		
		COD or TOC	NH_4-N or TKN	TN	COD or TOC	NH_4-N or TKN	TN
VF (sand) [a]	Unsaturated	97 [b]	99.9	14	8.6 [b]	2.1	0.4
	15 cm (drainage layer)	97 [b]	99.6	14	8.6 [b]	2.1	0.4
VF (coarse sand) [a]	Unsaturated	89 [b]	88	14	31.3 [b]	7.3	1.4
	15 cm (drainage layer)	91 [b]	81	32	31.9 [b]	6.7	3.3
Single-stage French VF[4]	20 cm	92	87 [c]	70	86	7.9 [c]	6.3
	40 cm	97	91 [c]	77	91	8.3 [c]	7.0
	20 cm + 100% Recirculation	98	96 [c]	83	63	5.8 [c]	5.0
	40 cm + 100% Recirculation	98	96 [c]	84	63	5.8 [c]	5.1

[a] Sand = 50 cm main layer of sand with gravel size 0.06 – 4 mm; Coarse sand = 50 cm main layer of coarse sand with gravel size 1 – 4 mm. Langergraber et al. (2004)
[b] TOC
[c] TKN
[d] Prigent *et al.* (2013)

Table 6.4 Performance of a reciprocating wetland compared to other conventional and intensified treatment wetland designs (calculated from Nivala *et al.*, 2013).

	Mass percent removal			Mass removal rate ($g/m^2 \cdot d$)		
	BOD_5	**NH_4-N**	**TN**	**BOD_5**	**NH_4-N**	**TN**
HF	81.1%	2.8%	23.2%	6.8	0.1	0.6
VF (sand)	99.5%	87.2%	27.6%	21.4	4.3	1.9
VF + aeration	99.4%	99.1%	44.6%	22.0	5.2	3.1
HF + aeration	99.9%	99.3%	40.6%	31.1	7.3	3.9
Reciprocating	99.3%	91.3%	72.3%	29.9	6.6	7.1

Reciprocating treatment wetlands will have higher investment and operation & maintenance costs due to the extra pumps and components required to move the water back and forth between cells. At a small scale, this can render the technology unreasonably complicated or too expensive for implementation. However, for larger systems, reciprocating or fill-and-drain technology can be a cost-effective technology choice. On sloped sites, the use of siphons can significantly decrease the ongoing operational costs of moving water from one cell to another (Austin and Nivala, 2009). The use of carefully designed drainage layers and high-volume, low-head pumps can also enable wastewater treatment over the long term that is competitive with other conventional wastewater treatment technologies (Behrends *et al.*, 2001).

6.6 AERATION

In recent decades, use of active aeration (e.g., an air pump connected to a subsurface network of air distribution pipes) has been applied to HF wetlands (Wallace, 2001; Higgins, 2003; Ouellet-Plamondon *et al.*, 2006; Maltais-Landry *et al.*, 2009; Butterworth *et al.*, 2016) as well as saturated VF wetlands (Murphy and Cooper, 2011; Wallace and Liner, 2011; van Oirschot *et al.*, 2015). In the UK, over 40 tertiary treatment HF wetlands have been successfully retrofitted with aeration systems to increase treatment capacity and extend asset life (Murphy *et al.*, 2016). The use of this wetland technology is not limited to domestic wastewater. Aerated wetlands have been successfully implemented for an array of complicated industrial wastewater streams as well, including (but not limited to) landfill leachate (Nivala *et al.*, 2007), airport deicing runoff (Murphy *et al.*, 2014), mine tailings (Higgins, 2003) and groundwater contaminated with petroleum compounds (Wallace and Kadlec, 2005). Aerated wetlands have also been used for treating CSO (Nivala *et al.*, 2014) and wastewater from visitor centres (Murphy *et al.*, 201[illegible]).

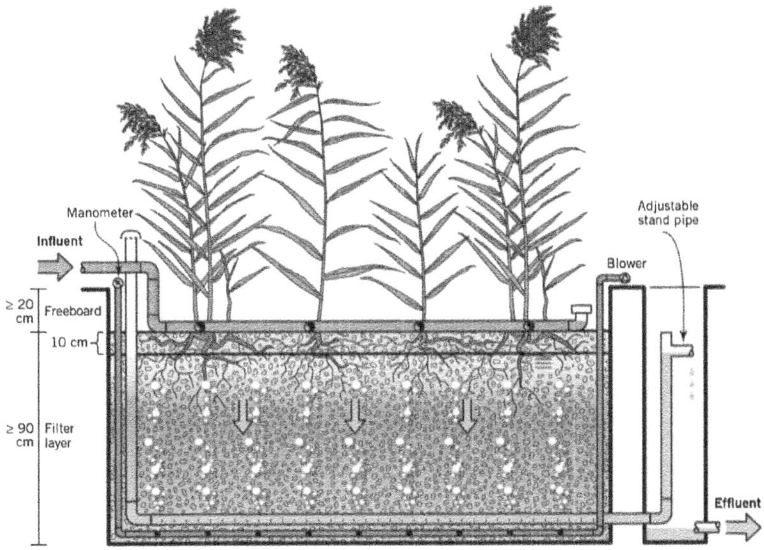

Figure 6.1 Schematic of an aerated wetland; VF.

Aerated wetlands are generally designed with a coarse gravel media and a saturated depth of at least 100 cm, and can be HF or saturated VF (Figure 6.1). In most aerated treatment wetlands, the water level in the wetland is regulated by the effluent standpipe and kept 5 – 10 cm below the surface of the gravel layer. In cold climates, a 15 – 20 cm insulating layer of well-decomposed mulch can be used to protect the system from freezing in the winter (Wallace and Nivala, 2005).

Aerated treatment wetlands are reported to have organic matter removal rates 10- to 100-fold higher than conventional treatment wetland designs (Nivala, 2012). As can be seen in Table 6.4 (previous section), aerated HF and aerated VF wetlands are both extremely efficient in removal of carbon and ammonium nitrogen, and remove TN more effectively than either traditional HF or VF designs. Intermittent aeration (multiple on-off aeration cycles per day) has been shown to improve TN removal in VF aerated wetlands (Fan *et al.*, 2013; Foladori *et al.*, 2013; Boog *et al.*, 2014).

7

Free water surface wetlands

7.1 INTRODUCTION AND APPLICATION

FWS wetlands were one of the first treatment wetland options to be implemented as they arguably mimic water purification processes within a natural wetland more than any other treatment wetland type. They are commonly used to treat non-point sources such as urban stormwater, agricultural runoff and metal-laden flows in addition to municipal wastewater (Vymazal, 2013). Due to a relatively low cost per unit area, they generally find their greatest application in high flow volume, low pollutant concentration situations. In domestic and municipal wastewater treatment applications, they are usually found downstream of other treatment units and are often considered a tertiary or polishing step. Aesthetic and habitat values are often as important to the design as water quality improvement.

The physical structure of a FWS wetland is as diverse as its potential application. They may be lined or unlined, constant or variable in depth, completely or partially vegetated, the vegetation can be emergent, submerged or floating and they can vary in size from a few square meters to multiple square kilometres. Yet there are several essential defining features. Water level is maintained above a rooting matrix of soil, sand or gravel that supports the growth of wetland plants that can survive continuously flooded conditions. Flow is horizontal but may take a circuitous path from inlet to outlet at a very low velocity (Figure 7.1). This chapter describes the typical FWS wetland treating low-strength municipal wastewater after most of influent TSS and COD or BOD_5 has been removed, e.g. following secondary treatment. Other applications of FWS wetlands include Floating Treatment Wetlands (Section 8.4), stormwater wetlands, permanently flooded bio-swales and wetlands employing submerged vegetation.

FWS wetlands depend on a diverse set of pollutant removal mechanisms, including physical sedimentation and pollutant degradation by chemical, microbial and photo pathways. More so than other treatment wetland variants, FWS wetlands simultaneously promote both aerobic and anaerobic processes and organic matter loading rates often determine which dominates. The rooting layer is largely anaerobic, especially after the system matures and a layer of detritus consisting of dead vegetation and incoming sediment establishes over it. The lower levels of the water column can range from anaerobic to aerobic depending on pollutant loading rates, depth of water column and distance from the flow entry point. The wetland should be designed so that the upper layers of the water column are always aerobic to prevent odour releases and promote death of pathogenic organisms. Virtually all redox-dependent reactions, including nitrification and denitrification, are possible in the FWS wetland due to this array of redox conditions. Open water areas allow sunlight to penetrate and enhance photo-degradation.

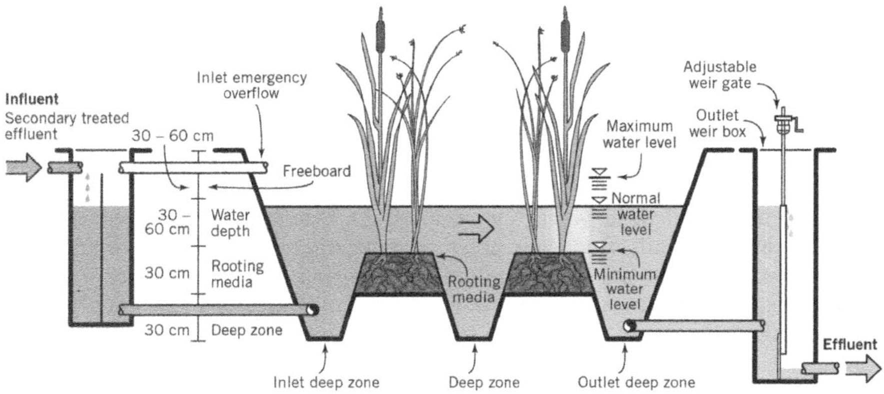

Figure 7.1 Overview of a FWS wetland.

Plant uptake in a FWS system plays a more significant role in nutrient removal than in other treatment wetland types. Plants also release small amounts of oxygen and organic carbon compounds into the rooting matrix, fuelling both aerobic and anoxic microbial processes. Most processes playing a role in FWS treatment wetlands are represented in Figure 7.2.

7.2 DESIGN AND WATER QUALITY TARGETS

While FWS wetlands have been used for secondary municipal wastewater treatment with only solids separation as prior treatment, they have fewer attachment sites for microbial biofilms and thus slower reaction rates compared to HF and VF wetlands, which are more cost effective for secondary treatment objectives. Enhanced tertiary nutrient removal, settling of solids such as algae generated in upstream treatment units, stabilisation of flow rate and pollutant concentration fluctuations, or recharge of groundwater with processed wastewater are more typical objectives for FWS wetlands. Provided there is sufficient area, the size can be increased to accommodate wildlife habitat and aesthetic considerations.

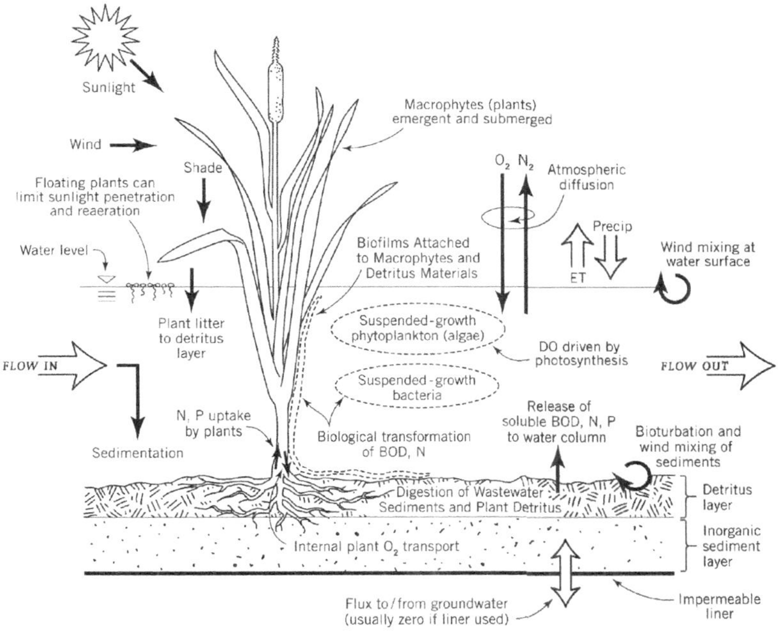

Figure 7.2 Major processes in FWS wetlands. Reprinted with permission from Wallace and Knight (2006).

Hydrology, hydraulics, and climatic considerations are important design factors. The influence of infiltration (if unlined), precipitation and ET on the water budget is enhanced due to the large surface area and combination of open water and emergent plants. Large surface area and low velocity increases the potential for hydraulic short-circuiting. Climatic limitations such as freezing and potentially significant infiltration and ET losses are exacerbated by water levels above the ground surface and long retention times. Minimisation of size for a given pollutant removal expectation, a typical first step in treatment wetland design, might not even be a consideration because FWS wetland water quality goals can be less important than aesthetic and other project goals. Yet the design engineer must ensure that the size determined from other project objectives is sufficient to achieve a given water quality target.

The diversity of potential applications, design objectives and treatment goals makes design of FWS simultaneously more complicated but also more forgiving compared to other treatment wetland options. They tend to be less optimised for removal of a specific pollutant or suite of pollutants and the designer relies more on background processes and process rates, not unlike those occurring in natural wetlands. As with all natural processes, complete removal of organic carbon and nutrients is not possible. In fact, wetlands will *produce* these compounds if influent values are less than background production rates. Consideration of background concentrations as characterised by C^* is much more important to performance prediction than degradation kinetic parameters when influent concentrations are near C^* values, because rates are slow regardless of first-order kinetic constants. Reported background concentrations in FWS wetlands are provided in Table 2.5.

Despite the similarities, there are important differences between FWS treatment wetlands and natural wetlands. FWS wetlands receive pollutant loads that are low compared to other (secondary) wastewater treatment systems, but are high compared to natural wetlands. Higher nutrient loading tends to increase plant biomass production and its eventual deposition as detritus. The rate of detritus accumulation can be accentuated because FWS wetlands are generally kept flooded, relying on anaerobic decomposition pathways even during periods of low water. This has several effects. Plant nutrient uptake is increased during periods of growth and is less likely to be re-released by decay of detritus. Thus, burial may be a significant seasonally dependent nutrient removal mechanism. Accumulating detritus increases oxygen demand in the root zone. Plants respond to higher oxygen stress by increasing oxygen transfer to roots, thus paradoxically aerobic processes in the rooting media may be enhanced even though the overall redox condition is highly reducing. The detritus is also an organic carbon source that can be used by denitrifying organisms to remove nitrate.

Hydrology

A general hydrologic budget is presented in Section 2.2. As with all treatment wetlands, it is important to isolate the system from runoff from surrounding land areas (unless treating runoff is a design objective). Suitable freeboard for precipitation falling onto the surface and an emergency overflow option should the design storm be exceeded must also be provided. The effect of the increased depth on the hydraulics of the outlet works should be considered because it is important to ensure that water level will return to normal within an appropriate drawdown period. Freeboard should also take into consideration the formation of a detritus layer and, when relevant, ice formation.

Losses due to infiltration (if unlined) and ET are generally more important than precipitation inputs because water level should be maintained above the rooting media to prevent rapid oxidation of the accumulated detritus layer. Infiltration from unlined systems must consider the initial infiltration rate and the slower rate anticipated once a detritus layer is established. Initial infiltration is highly dependent on soil properties, especially texture, but can be minimised by compaction or adding an expanding clay (such as bentonite) before adding the rooting media. When such actions are taken the decrease in infiltration rate due to detritus accumulation is minimised. Evapotranspiration losses can be significant in warm, arid climates and in summer in temperate climates. Its effect is to reduce outflow and to concentrate contaminants, thus pollutant degradation rates must exceed the rate at which ET concentrates pollutants. Due to open water above the soil surface, ET from FWS wetlands is often assumed to be similar to shallow open water bodies, and as a first approximation can be considered as 80% of readily available pan-evaporation rates. If calculated rates are a significant portion of the design flow, use of more precise methods is warranted.

Hydraulics

The depth of flow is limited to the maximum level a desired plant species can survive permanently under flooded conditions. An upper limit is approximately 60 cm but most systems are designed to have an average depth of around 30 cm. The depth can be increased in some internal locations to promote open water areas. Aligning open water areas in strips perpendicular to the flow direction can help limit hydraulic short-circuiting but may compromise wildlife and aesthetic considerations. The design depth must also account for hydrologic factors as discussed previously.

Flow velocities and average depths across a FWS wetland are such that flow is usually in the laminar or transitional range, thus hydraulic sizing cannot utilise turbulent-flow dependent equations such as Manning's Equation. However, the frictional drag created by a dense stand of emergent vegetation cannot be ignored even in laminar flow. Methods for estimating velocity in FWS wetlands are presented in Section 2.2. With a pre-selected average depth, the cross-sectional area of flow can be determined by use of Equations 2.1 and 2.2. The minimum surface area is then determined from water quality considerations as described in the next subsection.

Water quality

Because of their widespread use as a tertiary polishing step in municipal wastewater treatment, FWS wetlands typically target nutrient e.g. nitrate and/or phosphorus removal but have also been employed to reduce pathogenic organisms and/or suspended solids, specifically algae grown in upstream treatment units. The key design parameter is an appropriate wetland surface area to meet the discharge target or pollutant removal expectation. Some jurisdictions may prescribe a mass loading rate criterion e.g. (kg/ha·yr or $g/m^2 \cdot d$) but the modified first-order rate model is more typically utilised. Both volumetric and areal versions can be employed, but since the range in design depths for FWS wetlands is relatively narrow, the two methods yield similar results. In one case, determination of the appropriate surface area is direct. When the volumetric version is used, the calculated HRT is divided by the water depth in order to determine the required area. In either case, the length of the wetland is determined from the surface area after the width is determined of the cross-sectional area calculated from the hydraulic analysis. Reported areal rate coefficients are provided in Table 2.3.

Thermal balance

In cold climates, FWS wetlands must be protected from freezing in winter. One solution is to simply stop the use of the wetland in winter, either storing the water during this time for treatment the following summer or assuming the pollutant of concern (e.g., algae or pathogens) is not an issue in winter. Some jurisdictions prescribe this alternative. A second alternative is to raise the water level prior to the initiation of the freezing period in order to allow an ice layer to form. The water level is then lowered after the formation of the ice layer, which then enables the ice and air space between the ice and the water surface to provide enough thermal insulation to keep the water temperatures above freezing. It is important, however, to raise the water level before ice formation so that plant stems encased in ice are not uprooted as the water level rises. Determination of the proper increase in water level becomes an iterative process, balancing the rate of ice formation relative to the decreasing heat loss due to the increasing ice thickness for the worst-case temperature period. Wallace and Knight (2006) provide an example of how such a calculation can be made.

Other design considerations

Hydraulic short-circuiting and hydraulic dead-zones are common problems in large FWS wetlands. Alternating vegetation and open water zones across the flow path can help but inlet and outlet works designed to distribute and collect the flow across the entire width are essential components. It is also possible to break the required wetland area into a set of smaller, more manageable cells. Cells can be arranged in series or parallel. In either case, the hydraulic inefficiencies of any one cell are minimised.

Different wetland plant species have different stand densities, growth and senescence patterns, oxygen transfer potential, detritus accumulation rates and a host of other factors influencing bio-geochemical processes in wetlands. They also have varying flooding tolerances, which influences design depth. Moreover, an established wetland, especially a FWS wetland, will be colonised by species which were not originally planted. While species selection likely plays a role in performance (Brisson and Chazarenc, 2009; Taylor *et al.*, 2011), there is simply insufficient information to make plant selection a design criteria even if water quality performance is the only concern. The best option is to plant a FWS wetland with plants that are known to establish and grow well in the region and to consider plant diversity for its influence on habitat improvement objectives and insurance against catastrophic stand die-off. Commonly used plants include *Phragmites*, *Typha* and *Schoenoplectus* genera in temperate climates, and plants such as *Canna* or *Arundo* in tropical climates. The media in which plants will be rooted can also influence performance, but in most cases, native soil is used. If a liner is used, a typical plant media depth is 30 cm, but this layer gradually thickens over time as the detritus layer decomposes and collects on the bottom of the bed. Performance of the

FWS wetlands typically improves over the first few years as plants establish thick stands and the detritus layer accumulates.

7.3 OPERATION AND MAINTENANCE

FWS wetlands need very little maintenance under normal operating conditions. Periodic inspection of inlet and outlet works and plant health is advisable. Plants that are subjected to oxygen stress tend to concentrate roots closer to the surface, making them less tolerant of periodic deep-water conditions and more susceptible to lodging, thus complete submergence and death. The typical large scale of FWS wetlands makes them susceptible to wave action, which can exacerbate plant lodging and increase the potential of wind-induced bank erosion.

7.4 CASE STUDY

Wetland Alhagen is a FWS wetland in Sweden that was constructed during 1997 and became fully operational in 1998. One of the main objectives was to reduce the nitrogen load to the Baltic Sea from the wastewater treatment plant in Nynäshamn town, which was at that time equipped with mechanical and chemical treatment. As the wetland was the only biological wastewater treatment, it was designed to promote both nitrification and denitrification and was initially operated during the period April – December each year. During 2001, an SBR was added to the treatment plant and from the autumn 2002, the inflow to the wetland has been mechanically, biologically and chemically treated.

The wetland area covers 28 ha wet treatment area, with a retention time of approximately 14 days. It is situated in a valley with previous arable land on clay soils, though part of the wetland is a natural mire. Treated wastewater is pumped to the wetland inlet and flows by gravity through the system. In the first part of the wetland, wastewater is treated, and in the lower part the treated wastewater is mixed with settled stormwater from circa 200 ha (Figure 7.3). The annual load of stormwater is around 150,000 m3, or < 8% of the hydraulic load.

In the first part, wastewater flows through small elongated wetlands in series, originally intended for settling of sludge, and is then alternatively discharged to the West or East wetland with predominantly emergent macrophytes such as *Phragmites australis, Typha sp., Scirpus sylvaticus* and *Carex riparia*. Those are loaded every three to four days and operated in a tidal mode to improve oxygenation and achieve good hydraulic efficiency. Water from both those sections is collected in one elongated wetland and forwarded to the wetland Stordammen, dominated by emergent macrophytes and also operated in a tidal mode (Figure 7.3). From here, water is discharged twice a week to an overland flow area, Skålpussefallet, covered by *Typha latifolia* and forage grasses from which it is collected in a pond with mainly submerged plants and is then mixed with stormwater. This mixed water flows into two final shallow wetlands with emergent plants (*P. australis* and other species).

Figure 7.3 Wetland Alhagen with the first inlet wetlands and the alternately loaded West wetland (left), East wetland (background) and Inlet wetland (left of car) with bypass channel for stormwater between the pathways. Photo courtesy of Christer Svedin.

The N treatment performance has been stable and satisfactory during the entire operation period. For the period presented here, the mean total N removal was 61% for the period 2003–2009 (Table 7.1), when the concentration of NH_4^+-N in the inflow was 16 mg l^{-1} compared to 37 mg l^{-1} in the preceding three years. There is a strong seasonal variation in N removal from a low ca 30% during Feb – Mar to around 80% or higher during the warmer months. Apart from the N removal, the wetland functions as an efficient polishing step with respect to total phosphorous and BOD_7 with concentrations at a third of the allowed discharge levels, which is highly beneficial for the recipient Baltic Sea. Currently, the municipality staff is working on a project to optimise the combined operation of the SBR plant and the wetland to achieve as efficient treatment as possible.

Table 7.1 Wastewater characteristics and treatment results for wetland Alhagen, Nynäshamn for the 1999 – 2001 and 2002 – 2009.

	HLR (mm/d)		**TN (mg/L)**		**TP (mg/L)**		**BOD_7 (mg/L)**		**TN load (kg/ha·yr)**		**TP load (kg/ha·yr)**	
	99-02	**03-09**	**99-02**	**03-09**	**99-02**	**03-09**	**99-02**	**03-09**	**99-02**	**03-09**	**99-02**	**03-09**
In	16	0.019	37	22	0.39	0.30	35	9.7	1,607	1,506	17	20
Out			11	6.8	0.10	0.06	3.9	3.0	500	590		
Removed									*1,107*	*916*	*12*	*15*

8

Other applications

8.1 ZERO-DISCHARGE WETLANDS

In zero-discharge systems, as the name indicates, there is no discharge of treated wastewater. The system is designed to allow all the influent to be released to the atmosphere by plant ET. They rely upon actual ET rates that exceed precipitation rates on an annual basis. Principal features of these systems are zero discharge and nutrient removal via harvested plant biomass. Zero-discharge systems have been developed and implemented mainly in Denmark and Ireland (Gregersen *et al.*, 2003; Brix and Arias, 2005; Environmental Protection Agency Ireland, 2010; O'Hogain *et al.*, 2010), accentuating the fact that annual potential ET rates need not exceed annual precipitation for the systems to function properly. Because of their rapid growth and high ET rate, willows (*Salix viminalis L.*) are most commonly used for these types of systems.

The total annual water loss from zero-discharge willow systems can be assumed to be higher than the potential ET at the location as determined by climatic parameters. In small treatment wetlands, the vegetation experiences enhanced evaporation from the "oasis effect", resulting from warmer and dry air flowing across the plant canopy area of plants. In addition, there is also the "clothes-line effect", where the vegetation height is greater than that of the surroundings and may increase evaporative loss. Therefore, ET from isolated expanses, on a per unit area basis, may be significantly greater than the calculated potential ET (Kadlec and Wallace, 2009).

Zero-discharge systems can meet even the most stringent requirements because there is no outflow (Figure 8.1). According to Brix and Arias (2005), the main characteristics of the zero-discharge willow systems in Denmark can be summarised as follows:

- For a single household system (5 PE), the sewage must be pre-treated in a two or three-chamber sedimentation tank with a minimum volume of 2 m^3.
- Closed willow systems are generally constructed with a width of 8 m, a minimum depth of 1.5 m, and with 45-degree slopes on the sides.

- The total annual water loss from the systems is assumed to be 2.5 times the potential ET at the location as determined by climatic parameters.
- The required system area is determined by the amount of wastewater, the average annual precipitation, and the potential ET at the location of the system.
- The bed is enclosed by a watertight membrane and wastewater is distributed underground within the system by a level-controlled pump.
- A drainage pipe is placed in the bottom of the bed. The pipe can be used to empty water from the bed if salt accumulates after some years.
- One third or one half of the willows are harvested every year to keep the willows in a young and healthy state with high transpiration rates.

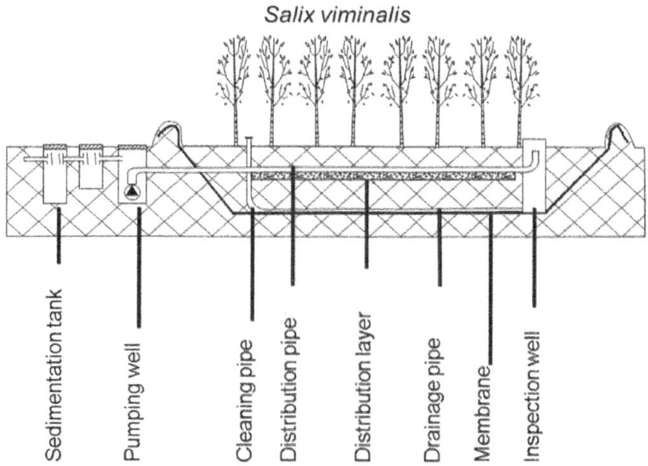

Figure 8.1 Schematic of a zero-discharge willow system. Reprinted with permission from Brix and Arias (2005).

Willow systems with soil infiltration are dimensioned in the same way as closed willow systems. The willows will evaporate all wastewater during the growing season, but during winter some wastewater will infiltrate into the soil. Willow systems have recently been piloted in extreme cold climate conditions in Mongolia with promising results (Khurelbaatar *et al.*, 2017).

Contrary to the Danish system, the zero-effluent system developed in Ireland is a combination between treatment wetlands and willow beds (O'Hogain *et al.*, 2010). The Irish system comprises treatment of the wastewater with VF and HF wetlands before the treated wastewater is discharged into a willow bed, which is designed without an impermeable liner. During the two-year investigation period of the study, no effluent from the willow bed occurred (O'Hogain *et al.*, 2010).

8.2 COMBINED SEWER OVERFLOW TREATMENT WETLANDS

Combined sewer systems are designed to transport stormwater surface runoff in addition to the dry weather flows. The treatment plant is designed to process the combined flow up to a predetermined maximum and, in many places, hydraulic loads greater than the design flow are discharged directly into receiving water bodies with minimal treatment (screening, sedimentation), or with no treatment at all. Management of CSO is a challenge for many communities. Treatment wetlands are considered one of the green infrastructure solutions to CSO treatment (Meyer *et al.*, 2013; Tao *et al.*, 2014).

VF wetlands are the preferred wetland type to treat CSO and are widely used (often in combination with other wetland types). In Germany, these systems have been developed and modified since the 1990s and a national design guideline was released in 2005 (DWA, 2005). The German design guidelines have been adapted recently in France and Italy (Meyer *et al.*, 2013). A few CSO treatment wetland systems of different configurations have also been applied in the USA (Tao *et al.*, 2014) and in the UK (Green *et al.*, 1999) since the 1990s. Ávila *et al.* (2013) proposed the combination of a VF wetland, HF wetland and FWS wetland in series to treat CSO in Spain.

Main treatment objectives for CSO treatment wetlands are: (1) detention and reduction of peak flows, (2) reduction of suspended solids by filtration and (3) reduction of soluble and particulate pollutants by adsorption and subsequent biological degradation (Dittmer *et al.*, 2005). To reach the first objective throttle valves are applied to limit the maximum effluent flow rate and therefore the flow velocity in the filter itself. Due to the limited effluent flow, water accumulates over the bed and a retention basin is required on top of the VF filter.

Fundamental differences between wetlands for wastewater exclusively and CSO treatments are the hydraulic loading regime and the quality parameters of the inflow. In the CSO treatment, the succession of loading events and dry periods is characterised by the stochastic nature of rainfall and the runoff behaviour of the catchment area. Extreme cases involve a permanent loading for weeks on the one hand, and several months without any loading event on the other. HLRs (inflow/filter area) show high variability ranging from mean values of 0.02 mm/s for less intensive rain events, up to peak flows of 1 mm/s during intensive storm events.

CSO flows are generally less concentrated than wastewater. Organic matter predominantly occurs in particulate form, which can mainly be attributed to remobilisation of sewer sediments. As this effect is strongly related to the flow rate, TSS concentrations, as well activated sludge solute/particulate ratio of pollutants, are highly variable within the course of a rain event.

Figure 8.2 shows simplified system sketches for CSO treatment wetland designs in Germany, France and Italy (Meyer *et al.*, 2013). The German design (Figure 8.2A) consists of CSO tanks with VF filter beds in series. The main layer of the VF filter has depth of 0.75 m and consists of fine sand (grain size 0 – 2 mm, minimum hydraulic conductivity 10^{-4} m/s). The retention basin has a height of 1 m. Outflow rates of the VF filter are limited to 0.01-0.03 L/(m^2·s). The French design (Figure 8.2B) is based on French VF wetlands, the two VF beds are loaded alternately. During very high flows (e.g. more than five – six times the dry weather flow), the second bed is also loaded. The Italian design (Figure 8.2C) focuses on the treatment of the first flush, i.e. the first 50 cm of rain. This results in smaller VF beds. After the first flush is treated, CSO is diverted into a FWS wetland.

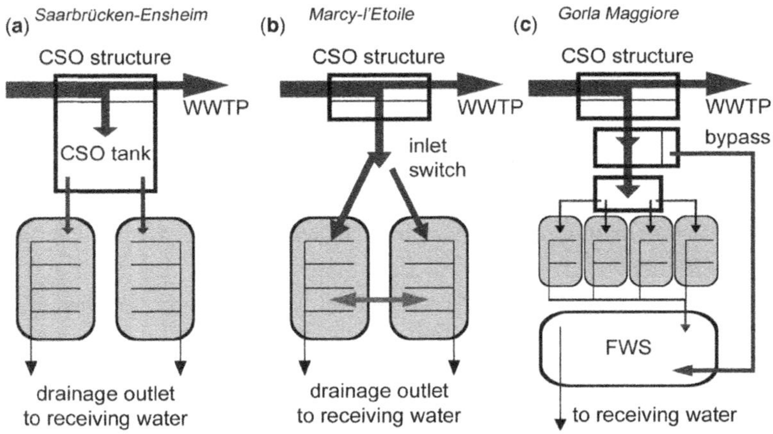

Figure 8.2 Simplified system sketch for CSO treatment wetland design in (A) Germany, (B) France and (C) Italy. Reprinted with permission from Meyer *et al.* (2013).

The main challenges in determining appropriate dimensions are common for all CSO treatment wetlands: 1) predicted flow rates of CSOs are derived from numerical simulations and thus of high uncertainty, and 2) the stochastic nature of rain events. According to Meyer *et al.* (2013), robustness in design can be improved in a number of ways: either by maintaining a permanent water layer as insurance against plant water stress, by decreasing the filter surface area for enhanced sediment distribution and/or by using at least two filter beds with alternating loading (to ensure rest periods and avoid clogging). The addition of FWS wetlands increases ecosystem services of biodiversity increases and recreational activities (Liquette *et al.*, 2016; Masi *et al.*, 2017b).

8.3 SLUDGE TREATMENT WETLANDS

Many types of conventional municipal wastewater treatment technologies (e.g. activated sludge) produce large quantities of sewage sludge. In general, sludge is produced during wastewater treatment in a liquid form, typically containing 0.5 to 5% Total Solids (TS) content. Per capita sludge production varies by country and ranges from 0.1 to 30 kg TS/PE·yr. Most of the components of the sludge generated are organic. Therefore, the volatile solids (VS) content of the sludge is quite high, generally ranging from 75% to 80% TS (von Sperling and Gonçalves, 2007). Besides treating waste activated sludge, sludge treatment wetlands are also suitable for treating faecal sludge from cesspits and also septic tank sludge. These systems are also referred to as "*planted drying beds*" and "*sludge mineralisation beds*" and experience exists in developing countries such as Thailand, Cameroon, Senegal and Brazil. Typical sludge loading rates for planted drying beds are 200 – 250 kg TS/m^2·yr (Strande *et al.*, 2014).

The description provided herein refers only to sludge treatment wetlands for treatment of excess sludge from municipal wastewater treatment plants. The main objectives of sludge treatment are to reduce its volume (by applying dewatering or thickening strategies) and decrease its reactivity (sludge stabilisation). The selected sludge treatment process or technology depends upon the final reuse or disposal strategy and the quantity of sludge to be treated. Accordingly, most widely applied sludge management strategies are based on a centrifugation or filtration process (followed in some cases by thermal treatment). However, dewatering strategies that demand less energy such as water evaporation, ET and percolation are also applied. Conventional mechanised sludge treatment strategies are costly and energy demanding, which may lead to economic infeasibility for small communities.

Sludge treatment wetlands (also known as *sludge drying reed beds*) consist of shallow beds filled with a gravel layer and planted with emergent macrophytes such as *Phragmites australis* (common reed). Fresh sludge is spread and stored on the surface of the beds where water is removed from the sludge via ET. In comparison to common mechanical dewatering technologies (such as centrifugation), sludge treatment wetlands have low energy requirements and lower O&M maintenance costs. Although there are no standardised design or operation guidelines for sludge treatment wetlands, there is long-term experience (> 20 years) with the technology (Nielsen, 2012). Full-scale facilities have been designed and operated for communities ranging from 400 to 125,000 PE depending on the country surveyed, and the specific area requirements range from 1.5 to 4 PE/m^2 (DeMaeseneer, 1997). A general schematic of a sludge treatment wetland is shown in Figure 8.3.

Sludge treatment wetland systems contain multiple cells that are operated in parallel. Sludge is batch-loaded onto the surface of one of the beds. After loading one cell for a few days, the bed is left to rest from a few days to a few weeks while the next load is directed to another cell, resulting in a series of loading and resting periods for each bed. The resting time depends upon treatment capacity, weather conditions, age of the system, dry matter content of the applied sludge and thickness of the accumulated sludge layer (Nielsen, 2003). During the resting period, sludge accumulated on the surface of the bed undergoes a drying process, which results in a thin layer of dried material that is subsequently cracked by the movement of the *Phragmites* stems. The cracks caused by the plants help to support air movement into the sludge layer (which enhances sludge stabilisation) as well as maintains the hydraulic conductivity of the accumulated sludge. The actual number of parallel beds per facility depends upon the treatment capacity of the facility and can range from two to 25 (Nielsen and Willoughby, 2005; Uggetti *et al.*, 2010).

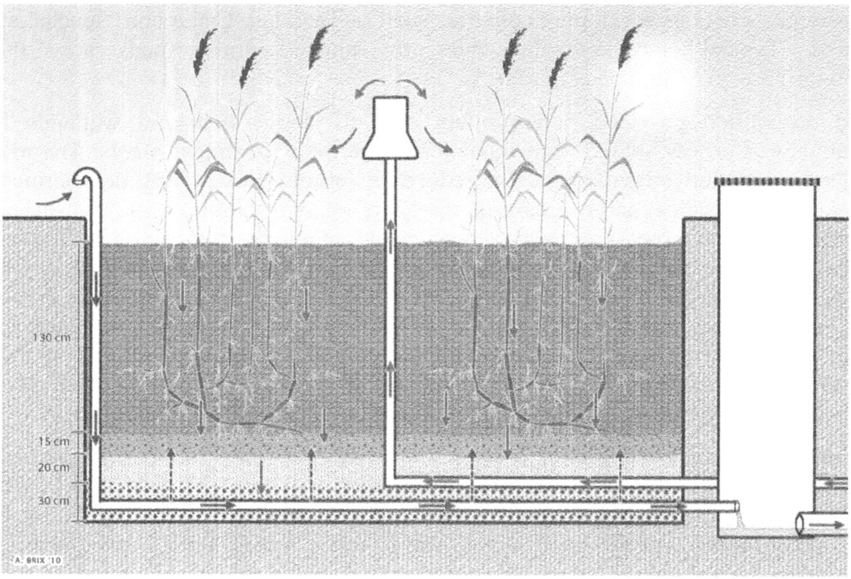

Figure 8.3 Schematic of a sludge treatment wetland. Reprinted with permission from Brix (2017).

The treatment cells are generally rectangular and constructed of concrete or in soil, the latter requiring an impermeable liner to prevent leaching. The cells are constructed with a minimum slope of 1% and any water that is not transpired by the plants is collected in perforated pipes at the bottom of the cell and returned to the head works.

One of the main operational parameters of sludge treatment wetlands is the areal loading rate. Maximum loading rates are in the range of 30 to 60 kg TS/m^2 yr (Nielsen, 2012). However, under warmer climates as in the Mediterranean basin, loads up to 90 kg TS/m^2·yr can be applied (Stefanakis and Tsihrintzis, 2012a). Sludge loading during the start-up phase (which can take from a few months up to three years) should be lower than these maximum loads in order to support proper plant establishment (Nielsen, 2003). The freeboard of the basin is also of importance since it determines the total life span of the system. Accordingly, most sludge treatment wetlands are approximately 2.5 meters deep (0.6 to 0.7 m of filter medium and 1.7 to 1.8 m for sludge accumulation). It is recommended that the bed should be able to store and treat at least 1.0 m of sludge accumulation. Assuming an accumulation rate of approximately 10 cm/yr, this results in a minimum period between emptying of 8 to 10 years (Nielsen, 2012).

The TS content of dewatered sludge is around 20 to 40%. This is comparable to the dewatering capacity of conventional technologies such as centrifuges (20 to 30% TS), belt filters (15 to 25% TS) and vacuum filters (20 to 35% TS). Reduction of VS in sludge treatment wetlands generally ranges between 25 to 30%, which results in VS concentrations in the stabilised sludge ranging from 40 to 50% VS. VS removal is dependent on the quality of the influent sludge and is thus higher for AS and lower for anaerobic digestion sludge and septage.

8.4 FLOATING TREATMENT WETLANDS

Floating treatment wetlands represent a group of wetland technologies where a buoyant structure is used to grow emergent macrophytes on a pond, lake, river or similar water body. Although the early applications of floating treatment wetlands date back to the early 1990s, the development and implementation of the technology has grown rapidly in recent decades. Floating treatment wetlands lend themselves to providing ancillary benefits, such as enhancement of habitat and aesthetic values. Applications of floating treatment wetlands include:

- Stormwater
- Polluted water canals
- CSO
- Sewage
- Acid mine drainage
- Animal production effluent
- Water supply reservoirs

A floating treatment wetland consists of emergent wetland vegetation growing on a mat or structure that floats on the surface of a pond-like water body (Headley and Tanner, 2012). The plant stems remain primarily above the water surface, while their roots grow down through the buoyant structure and hang in the water column (Figure 8.4). The plants grow essentially in a hydroponic manner, taking the majority of their nutrition directly from the water column. A hanging network of roots, rhizomes and attached biofilm forms beneath the floating mat, which provides a biologically active surface area for biochemical processes to occur as well as physical processes such as filtering and entrapment of particulates. Thus, a general design objective is often to maximise the contact between the root-biofilm network and the polluted water passing through the system. The depth of root penetration will depend largely on the plant species used and the physio-chemical conditions that develop in the water column below the floating plants.

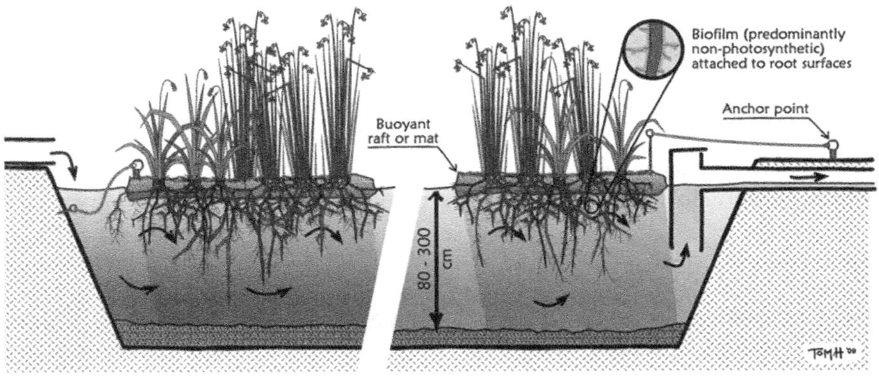

Figure 8.4 Schematic of a typical floating treatment wetland system. Reprinted with permission from Headley and Tanner (2012).

Headley and Tanner (2012) provide a comprehensive review of the available floating treatment wetland performance data, including mass removal rates and areal *k*-rates. Although data from long-term, full-scale applications are somewhat limited, data from mesocosm and pilot studies indicate that floating treatment wetlands can significantly improve the performance of pond systems and provide similar or better performance than surface flow wetlands for removal of organic matter, suspended solids, nutrients and metals. However, further studies on the growing number of full-scale applications are needed to verify the treatment performance and translate these observations into rigorous design methodologies.

Naturally occurring floating marshes exist in many parts of the world, where the right combination of factors has led to their development. However, the natural processes that lead to autonomous formation of large self-buoyant mats of emergent macrophytes are relatively slow and difficult to control. Thus, floating treatment wetlands are typically constructed using a floating raft or mat structure onto which suitable emergent macrophytes are planted. These are often modular in design, so that smaller, manageable individual units are joined together as needed to form larger rafts. A range of materials has been used for creating the floating rafts, including bamboo or plastic pipes and fabricated buoyant plastic mats made specifically for supporting floating wetland plants in a pond environment. The various construction techniques vary in cost, durability and effectiveness.

8.5 MICROBIAL FUEL CELL TREATMENT WETLANDS

Microbial Fuel Cells (MFCs) are bioelectrochemical systems that generate current by means of electrochemically active microorganisms as catalysts (Logan *et al.*, 2006). In a MFC, organic substrates are oxidised by exoelectrogenic bacteria and the electrons are transferred to the anode from where they flow through a conductive material and a resistor to a higher redox electron acceptor, such as oxygen, at the cathode (Logan *et al.*, 2006). In order for a MFC to work, there has to be a source of organic matter at the anode, a suitable electron acceptor at the cathode, and enough redox gradient (electromotive force) between the cathode and the anode. A redox gradient between the anode and the cathode can be obtained by either implementing a Proton Exchange Membrane (PEM) between electrodes (Figure 8.5) or by taking advantage of natural redox gradients encountered in aquatic environments (such as rice paddy fields, ponds and wetlands among others). MFCs that operate without a PEM are generally known as benthic or sediment MFCs. The compounds that are oxidised at the anode of a MFC are mainly simple carbohydrates such as glucose or acetate that are already present in the environment, or which are obtained from the microbial degradation of complex organic substrates such as organic sediments or wastewater (Rabaey and Verstraete, 2005). MFCs are therefore a technology to harvest energy directly from wastewater in the form of electricity (Lefebvre *et al.*, 2011).

MFCs can be implemented in HF wetlands because of the abundance of organic matter available in the subsurface environment and the presence of a naturally generated redox gradient between the upper and deeper layers of the treatment bed (Corbella *et al.*, 2014). The implementation of MFCs in HF wetlands (so-called MFC treatment wetlands) is very recent and most knowledge is from laboratory-scale studies focused mainly on the quantification of the amount of energy that can be harvested (Doherty *et al.*, 2015). However, some studies report that implementation of MFCs in treatment wetlands may provide a higher degree of treatment and might also be suitable for monitoring treatment performance (Fang *et al.*, 2013) or monitoring clogging in HF wetlands (Corbella *et al.*, 2016a).

Currently available information on cell architecture, materials and operational modes of MFC treatment wetlands is rather limited. Figure 8.6 shows current available strategies for treatment wetlands operated as vertical flow MFC (adapted from Doherty *et al.*, 2015) and Figure 8.6 (bottom), shows treatment wetlands operated under HF MFCs. Studies dealing with these aspects have only recently been published (Yadav *et al.*, 2012; Fang *et al.*, 2013; Liu *et al.*, 2013; Zhao *et al.*, 2013). The most common material for electrodes are graphite-based materials (either in the form of rods, granules or plates). However, a gravel-based anode with a metal mesh electron collector ("cathode") has been shown to produce good results in terms of treatment efficiency (Corbella and Puigagut, 2015). In order to maximise current production, most studies have incorporated unrealistic operation modes (such as up-flow batch feed loading regimes) which maximise the redox gradient. Thus, there is still a need for assessing the true extent of treatment efficiency improvement resulting from the MFC implementation in treatment wetlands.

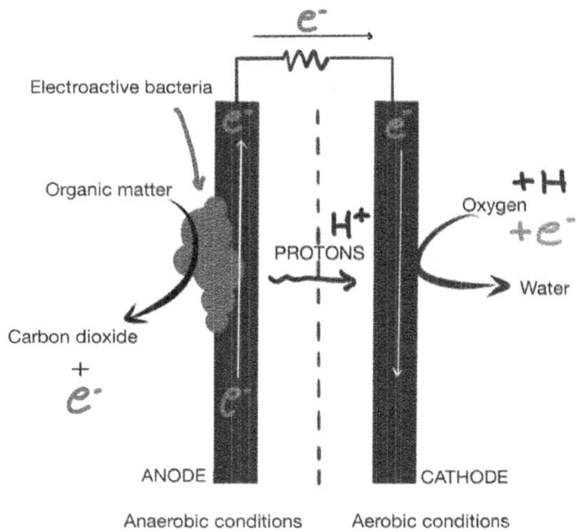

Figure 8.5 General schematic of a microbial fuel cell with a proton exchange membrane.

One of the main goals of coupling MFC and TWs is to generate an energy surplus while wastewater is being treated. Traditional treatment wetland designs consume very little energy (<0.1 kWh/m^3) (Kadlec and Wallace, 2009). Intensified wetlands often have slightly higher energy consumption rates (0.16 to 0.49 kWh/m^3) (Austin and Nivala, 2009; Kadlec and Wallace, 2009). Energy production in MFC wetlands ranges from only $2 - 45\ mW/m^2$ depending on the substrate, materials and configuration applied. As a result, MFC wetlands may not cover even 1% of the overall energy required to run, for example, an aerated wetland (Corbella *et al.*, 2016b). Therefore, the implementation of MFCs in treatment wetlands is not an advantageous strategy in terms of energy production alone.

MFC wetlands have been shown to be more effective in organic matter removal than traditional HF wetlands. The mechanism for this treatment efficiency improvement is not clear, but seems to be related to a higher biomass activity or a sulphate regeneration process (Li and Yu, 2015). MFC wetlands have been shown to remove approximately 10% more COD than traditional HF wetlands (Fang *et al.*, 2013). Laboratory-scale MFC wetlands fed with domestic wastewater and operated under batch mode have also been shown to increase not only the removal of organic matter (to a maximum of 30% higher than control systems), but also the removal of ammonia (approximately 20% higher removal than control systems) and phosphate (about 50% higher removal) (Corbella and Puigagut, 2015). Enhanced phosphorus removal seems to be related to the formation of phosphorus precipitates at the cathode of the MFC due to locally higher pH conditions. However, phosphorus removal is likely short-term in nature and the results from bench- or laboratory-scale experiments may be influenced by the scale and duration of the experiments. Further research is required in order to determine whether sustainable phosphorus removal is possible in full-scale systems.

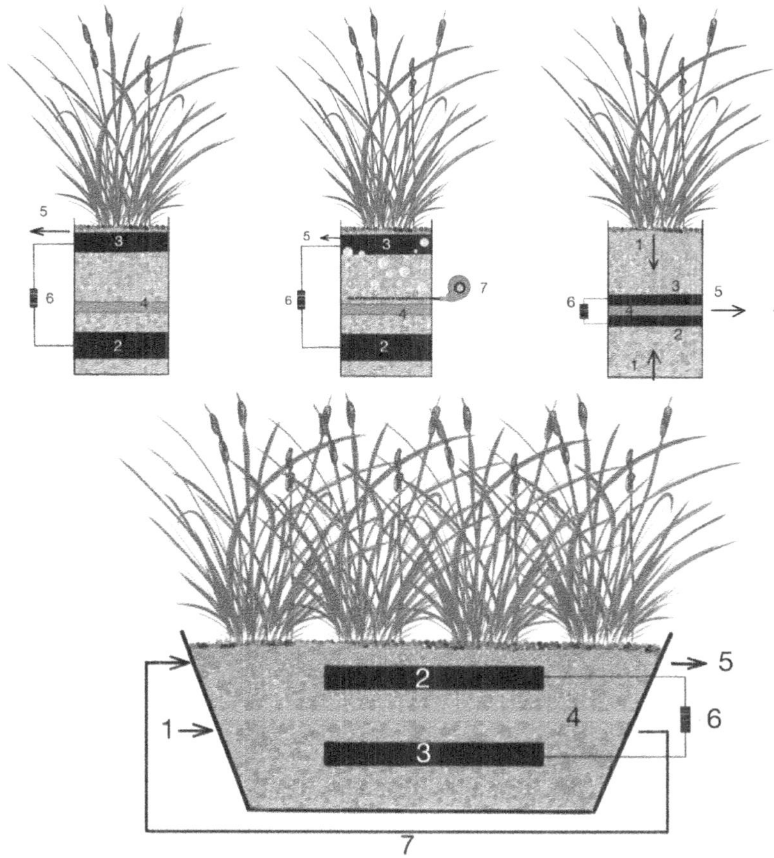

Figure 8.6 Schematics of most common design of laboratory-scale constructed wetlands operated as MFCs; top: vertical (up/down) flow modes; bottom: HF mode. *Note*: 1: influent; 2: anode; 3: cathode; 4: separation layer; 5: effluent; 6: external resistance; 7: aerator. Adapted from Doherty *et al.* (2015).

9

Additional aspects

9.1 PROCESS-BASED MODELS

During the last few decades, a number of mathematical models have been developed to describe the removal of pollutants in treatment wetlands. Most of the models published to date are simple "black-box" models where data from measurements are used to derive the model equations. Examples of "black-box" models include:

- Correlation models that correlate influent and effluent concentration,
- First-order rate equations, and
- More sophisticated correlation models such as artificial neural networks and fuzzy logic models.

The first two model types are typically used for designing HF and FWS wetlands (Chapters 3 and 7). As data from experiments are used for deriving the model parameters, good designs can be obtained only when the parameters have been determined from treatment wetlands operating under similar conditions (climatic conditions, wastewater composition, porous filter material, plant species, etc.).

In contrast, the governing equations in process-based models are derived from the processes occurring in wetlands. These types of models have various degrees of complexity and are predominantly based on balance equations (e.g. for energy, mass, charge). Experimental data are thus not required for deriving the model equations but are rather used for calibration and validation of the model. During the calibration and validation of process-based models, experimental data are compared with simulation results.

In order to describe the various processes occurring in treatment wetlands, a number of different sub-models must be considered:

- The *flow model*, describing water flow in the treatment wetland is of utmost importance.

- The *transport model*, describing transport of constituents, as well as adsorption and desorption processes.
- The *biokinetic model*, describing biochemical transformation and degradation processes.
- The *plant model*, describing processes such as growth, decay and decomposition of plants as well as uptake and release of substances such as nutrients, organic carbon and oxygen from roots.
- The *clogging model*, describing clogging processes, i.e., transport and deposition of suspended particulate matter and bacterial and plant growth that may reduce the hydraulic capacity or conductivity of the filter medium.

Process-based models have been developed mainly for HF and VF wetlands. The most advanced models available include:

- The HYDRUS Wetland Module (Langergraber and Šimůnek, 2012) implemented in the HYDRUS simulation software (Šimůnek et al., 2012), and
- BIO_PORE (Samsó and Garcia, 2013) implemented in the COMSOL Multiphysics™ platform.

Both the HYDRUS Wetland Module and BIO_PORE use multi-component biokinetic models, i.e. CW2D (Langergraber and Simunek, 2005) and/or CWM1 (Langergraber *et al.*, 2009b), respectively. Table 9.1 lists processes and components described by the biokinetic models CW2D and CWM1 as well as typical application of these models. Table 9.2 compares sub-models implemented in the HYDRUS Wetland Module and BIO_PORE.

Table 9.1 Processes and components described by (a number of components/processes in parentheses) and applications of the biokinetic models CW2D and CWM1.

Biokinetic model	**CW2D**	**CWM1**
Processes	Aerobic and anoxic (9)	Aerobic, anoxic, and anaerobic (17)
Components	Oxygen, organic matter, nitrogen, and phosphorus (12)	Oxygen, organic matter, nitrogen, and sulphur (16)
Type of CW	VF wetlands Lightly loaded HF wetlands	VF and HF wetlands

Table 9.2 Comparison of sub-models in the HYDRUS Wetland Module and BIO_PORE.

Simulation tool	HYDRUS Wetland Module	BIO_PORE
Reference	Langergraber and Šimůnek (2012)	Samsó *et al.*, (2016)
Flow model	Richards' Equation (variably saturated flow)	Variable water table (saturated flow)
Transport model	Advection, dispersion, adsorption	Advection, dispersion, adsorption
Biokinetic model	CW2D + CWM1	CWM1
Influence of plants	ET, uptake and release of substances	ET, uptake and release of substances
Clogging model	Not included	Included

Experience with applying the existing simulation tools for treatment wetlands can be summarised as follows:

- Process-based models are great tools for understanding the processes in treatment wetlands in more detail, e.g. spatial distribution of bacteria and pollutants in the filter, development of clogging, and response to sudden loads (Langergraber and Šimůnek, 2012; Samsó *et al.*, 2016; Langergraber, 2017).
- Simulation studies often suffer from a lack of sufficient data. Thus, when process-based models shall be used, experiments must be planned with this in mind. Data requirements are different for different models, but careful planning of sampling frequency and analysed parameters is essential (Meyer *et al.*, 2015).
- Good calibration of the water flow model is a pre-requisite for achieving a good match between measured and simulated pollutant concentrations. If the water flow model is calibrated, good results can be obtained in most applications when using the standard parameter sets of the biokinetic CW2D and CWM1 models (Langergraber and Šimůnek, 2012).
- Influent fractionation (i.e., fractionation of influent COD and the N and P contents of different COD fractions) has a high impact on simulation results and thus is an essential part of calibrating reactive transport models. This is especially true when simulating CWs treating other than domestic wastewater.

One of the main obstacles for the wider use of available simulation tools is that they are rather complicated and difficult to use. Simplified, yet robust and reliable, models for the design of CWs need to be developed, such as RSF_Sim, which was developed to support the design of CWs treating CSO (Meyer and Dittmer, 2015).

9.2 MICROPOLLUTANTS

As water quality regulations become more stringent, wastewater treatment plants are being evaluated for more than just removal of organic matter (carbon) and nutrients (N, P). In recent years, the removal of additional classes of chemicals has become a topic of significant interest. Wastewater treatment plants are now also being evaluated for their ability to remove certain groups of chemicals, including pharmaceuticals, personal care products, steroid hormones, surfactants, industrial chemicals and pesticides (Table 9.3). These chemicals are collectively referred to as *micropollutants* or *emerging organic contaminants*. Nowadays, these chemicals have been defined as priority substances in regulations worldwide, e.g. in Annex II of Directive 2008/105/EC in the EU (European Commission, 2008), in the list of toxic and priority pollutants under the Clean Water Act in the USA (U.S. EPA, 2016), and in the Priority Existing Chemical Assessments in Australia (NICNAS, 2016), respectively.

Over the past two decades, researchers have investigated the occurrence of micropollutants in wastewater in order to discover if and how they are removed by different treatment systems. Conventional wastewater treatment systems (mainly AS systems, membrane bioreactors and advanced oxidation processes in particular), have been the object of a great number of studies (Verlicchi and Zambello, 2014). Given their diverse properties (e.g., hydrophobicity and biodegradability) and low concentrations, micropollutant removal in current wastewater treatment plants is commonly variable and often incomplete (Luo *et al.*, 2014).

Experience shows that treatment wetlands, although not designed for the specific purpose, can remove micropollutants. Li et al. (2014) and Verlicchi and Zambello (2014) reviewed the performance of treatment wetlands regarding removal of pharmaceuticals. In general, wetland systems perform quite well compared to conventional treatment technologies such as AS for most of the pharmaceuticals reported. Among the different wetland systems, VF wetlands showed the highest removal performance for most substances, followed by HF wetlands, then FWS wetlands (Table 9.4). A similar trend was reported in a review of treatment wetland performance for removal of personal care products (Verlicchi *et al.*, 2014).

Table 9.3 Categories and major sources of micropollutants in the aquatic environment (Luo *et al.*, 2014).

Category	Important subclasses	Major sources
Pharmaceuticals	Nonsteroidal anti-inflammatory drugs, lipid regulator, anticonvulsants, antibiotics, β-blockers, and stimulants	Domestic wastewater (from excretion) Hospital effluents Run-off from concentrated animal feeding operations and aquaculture
Personal care products	Fragrances, disinfectants, UV filters, and insect repellents	Domestic wastewater (from bathing, shaving, spraying, swimming, etc.)
Steroid hormones	Oestrogens	Domestic wastewater (from excretion) Run-off from concentrated animal feeding operations and aquaculture
Surfactants	Non-ionic surfactants	Domestic wastewater (from bathing, laundry, dishwashing, etc.) Industrial wastewater (from industrial cleaning discharges)
Industrial chemicals	Plasticisers, fire retardants	Domestic wastewater (by leaching out of the material)
Pesticides	Insecticides, herbicides and fungicides	Domestic wastewater (from improper cleaning, run-off from gardens, lawns and roadways, etc.) Agricultural runoff

In general, treatment wetlands remove a variety of micropollutants that are usually present in domestic wastewater. For many substances, treatment wetlands perform better compared to conventional technical treatment systems. This is mainly because biofilms in treatment wetlands have coexisting aerobic-anoxic-anaerobic microenvironments and that there is a long "sludge retention time" compared to conventional technical systems. Both factors facilitate degradation of more complex organic molecules such as micropollutants.

Table 9.4 Removal efficiencies (in %) of selected pharmaceutical compounds in treatment wetlands and activated sludge (AS). Data from Li et al. (2014) and Verlicchi and Zambello (2014).

Pharmaceutical compound		**Secondary treatment**				**Tertiary treatment**		
Group	***Substance***	**AS**	**VF**	**HF**	**FWS**	**VF**	**HF**	**FWS**
Analgesic/ anti-inflammatory	Diclofenac	35	63	27	28	79	8	58
	Ibuprofen	90	85	55	65	69	48	60
	Naproxen	75	84	60	50	42	14	52
Psychiatric Drugs	Carbamazepine	10	25	25	25	26	60	25
Stimulants	Caffeine	90	95	85	75	n.r.[a]	n.r.[a]	5

[a] n.r. = not reported

9.3 ECONOMIC ASSESSMENT

Economic assessment is based on two components: capital costs (which are related to land acquisition and construction of the system), and O&M costs (which are related to plant management and operation on a yearly basis). Additionally, annual costs are used to describe cost per year of owning, operating and maintaining an asset over its entire lifespan. For this, the net present values of capital costs and future reinvestment costs are calculated and divided by the lifetime of the asset.

The construction of a treatment wetland system is usually carried out using local labour and materials. Therefore, it is very difficult to generalise either capital or O&M costs since they are site-dependent and may vary even within the same area depending on market conditions. In general, the associated capital costs for the construction of a treatment wetland are within the same order of magnitude as conventional wastewater treatment technologies, but the O&M costs are much lower due to the mechanical simplicity of treatment wetlands (e.g., fewer moving parts) and the fact that treatment wetlands generally have lower electricity requirements than other conventional wastewater treatment technologies.

Capital costs

Costs associated with the construction of the treatment system can be divided into direct costs (land acquisition, earthwork, pipes and fittings, pumps, filter media, liner, and plants) and indirect costs (site evaluation, permitting, and start-up services).

- *Land acquisition.* Wetland technology has higher surface area requirements than conventional wastewater treatment technologies for comparable treatment objectives. Therefore, land acquisition is often the largest capital cost, especially in locations with high land valuations such as highly developed urban areas.

- *Earthwork*. The construction of a treatment wetland requires site excavation and grading in order to produce cells of appropriate dimension that are enclosed by earthen berms. The costs associated with earthwork are highly dependent on the topography of the site, the quality of soil and the availability of labour and equipment, which is driven by local market conditions.
- *Construction materials*. Liner, filter media and labour are the most important cost components of wetland construction. Again, the costs for construction materials are highly site specific and vary widely. For the liner, either clay or synthetic material (polyethylene) can be used. When using a clay layer, costs depend on whether suitable clay is available near the site, since transportation costs are a significant factor. Synthetic materials are much more expensive to purchase but are much easier to transport and install. Sand, gravel and coarse rock are the most widely used filter materials in treatment wetlands and the size gradations are usually rather specific. The cost of these materials also depends on the availability of appropriate gradations at the source or whether specific sieves must be used, and the transportation costs, associated with the distance between the source and wetland site. The cost of additional materials such as piping, pumps and plants are generally small compared to the costs for the liner, filter media and construction labour.
- *Indirect costs*. In addition to the direct costs already mentioned, indirect costs also occur. These include engineering-related costs such as conceptual design, final sizing, preparation of plans, as well as non-construction contractor costs such as insurance and construction surveying. Construction observation and start-up services, as well as contingency and escalation costs are indirect costs that must also be considered (Kadlec and Wallace, 2009).

In Europe, capital costs of small treatment wetlands (excluding costs for land) are in the same range as packaged treatment plants. This is mainly due to high labour costs, because more labour is required to construct a treatment wetland than, for example, a sequencing batch reactor (SBR) or similar kind of wastewater treatment system. It can be expected that in countries with lower labour costs, the capital cost of a treatment wetland is more favourable. However, this can be counteracted by high costs of the liner (that needs to be imported), and/or high costs of sand or gravel of good quality that is not locally available and has to be transported over a long distance.

Operation and maintenance costs

Wetland systems intrinsically have very low O&M costs. Main O&M cost components include energy to run pumps (when necessary), compliance monitoring, maintenance of access roads and berms, and replacement or repair of mechanical components. If the wetland system can be loaded via gravity, no external energy is required and thus no pumping costs occur. Costs for vegetation management depend on items such as harvesting and pest control. Current design guidelines for VF wetlands foresee cutting of plants and removal of litter every two to three years. In general, less labour is required for O&M of treatment wetlands compared to other wastewater treatment plants. Thus, O&M costs of treatment wetlands are lower than those for other conventional wastewater treatment technologies.

The annual O&M requirements and costs, as an example, are given for a single-stage VF wetland with design size of 8 PE for a farmhouse in Austria (as of 2015):

- 180 € per year for external sampling and analysis that is requested by the authorities.
- 120 € per year for removing the primary sludge from the septic tank.
- One hour per month for additional sampling and analysis (pH and ammonia nitrogen effluent concentration) and visual checking of screen and grit chamber (including cleaning if required).
- Five hours per year for vegetation maintenance, i.e. wetland plants are cut in fall, then put on surface of VF bed for insulation in winter and plant material is removed from surface in spring).

It must be noted that all wetlands will also eventually require the removal of accumulated solids. The time frame for this cleaning or refurbishment operation will depend on the wetland design, loading rates applied, and the correct maintenance of upstream processes and routine wetland operation. Typical refurbishment intervals can be in the order of a decade for most systems (VF, first stage French VF, and tertiary HF), with more lightly loaded systems in the order of 20+ years (secondary HF, tertiary FWS). The designer must consider the trade-offs in capital and O&M costs for the individual scenario, as shorter refurbishment intervals are sometimes more economical than building a bigger system (Dotro and Chazarenc, 2014).

9.4 ENVIRONMENTAL ASSESSMENT

The concept of environmental sustainability is complicated, but it is linked to economic, social and environmental aspects. Sustainability can be defined as "*...the development that meets the needs of the present generation without compromising the ability of future generations to meet their own needs*" (WCED, 1987).

Life Cycle Assessment (LCA) is one of the most widely used methods for environmental assessment. The LCA methodological principles are based on the ISO 14040 standard (International Organization for Standardization, 2006). Other environmental assessments exist, but this chapter considers only the LCA method. The different phases of LCA include:

- *Goal and scope definition.* Describing the functional unit used, the conceptual, geographical and temporal boundaries of the system, the type and extent of the impacts considered, the data necessary to characterise the system, and the limitations of the study.
- *Inventory analysis.* Collecting and analysing data in order to quantify the inputs and outputs of the system, corresponding to the use of resources (energy and raw materials) and to the release of emissions (air, water, soil) for the entire life cycle of the system. In the impact assessment phase, the emissions catalogued in the inventory analysis are translated into their potential effect in the environment.
- *Impact assessment.* Consists of: i) selection of impact categories, category indicators, and characterisation models; ii) the classification stage, where the inventory parameters are sorted and assigned to specific impact categories; and iii) impact measurement.
- *Interpretation.* Consists of crossing the information from the inventory and/or impact assessment phases to produce conclusions and recommendations. Accordingly, recommendations are produced after a sensitivity analysis of key LCA is performed.

LCA applied to treatment wetlands

Impact categories generally considered for LCA on treatment wetlands include: Acidification Potential, Global Warming Potential, Eutrophication Potential, Freshwater Aquatic Ecotoxicology, Abiotic resources Depletion Potential, and Ozone Layer Depletion (OD). In general terms, LCA analysis on wetland technology is conducted in order to determine the main aspects of construction or operation affecting a certain impact category, or to determine whether wetlands are a more sustainable alternative than conventional wastewater treatment technologies.

LCA analysis applied to treatment wetlands indicates that both construction (materials and civil works) and operation are important factors influencing most impact categories. Accordingly, operation and construction roughly represents 30% to 60% of the impact depending on the category considered (Flores Rosell, 2015). Materials and processes that have the greatest contribution include energy consumption, metals and plastics production and manufacturing, crushed gravel production and chlorine. Concerning global warming potential, direct emissions show a similar impact as operation and construction. Sludge treatment produces only a significant contribution to the eutrophication category (representing almost 50% of the impact). Water re-use (if applied) lowers the impact between 25 and 55% depending on the category considered (Flores Rosell, 2015). Results from an overall comparison between treatment wetlands and conventional AS technology show that the AS technology has 1.5 to 6 times higher impact than treatment wetlands depending on the category considered, mostly because of the higher energy requirements and reagents used during operation. LCA performed on HF and VF wetlands shows that, overall, VF wetlands produce half (or even less) environmental impacts than HF wetlands mainly due to better treatment efficiency and because VF wetlands are smaller and have lower greenhouse gas emissions (Fuchs *et al.*, 2011).

LCA comparing sludge treatment wetlands and conventional sludge treatment technologies in small communities (< 2,000 PE) shows that the transport of the sludge produces the highest impact of all considered categories (Uggetti *et al.*, 2011). For sludge treatment wetlands where sludge is managed on site, the biggest impact is caused by raw materials used during construction (gravel, concrete, etc.). If the sludge that is dewatered and stabilised with sludge treatment wetlands has to be transported off-site, the overall impact of sludge treatment wetlands is equivalent to that of conventional mechanical-based sludge treatment technologies such as centrifuges or filter bands (Uggetti *et al.*, 2011).

10

References

Akratos C.S., Tsihrintzis V.A. (2007) Effect of temperature, HRT, vegetation and porous media on removal efficiency of pilot scale horizontal subsurface flow constructed wetlands. Ecological Engineering 29(2):173-191.

Andersson J.L., Kallner Bastviken S., Tonderski K.S. (2005) Free water surface wetlands for wastewater treatment in Sweden- nitrogen and phosphorous removal. Water Science and Technology 51(9):39-46.

Arias L. (2013) PhD Thesis. Vertical flow constructed wetlands for the treatment of wastewater and stormwater from combined sewers. L'Institut National des Sciences Appliquées de Lyon, Civil Engineering. Lyon, France. http://theses.insa-lyon.fr/publication/2013ISAL0102/these.pdf.

Austin D., Nivala J. (2009) Energy requirements for nitrification and biological nitrogen removal in engineered wetlands. Ecological Engineering 35(2):184-192.

Austin D.C. (2006) Influence of cation exchange capacity (CEC) in a tidal flow, flood and drain wastewater treatment wetland. Ecological Engineering 28:35-43.

Ávila C., Salas J.J., Martín I., Aragón C., García J. (2013) Integrated treatment of combined sewer wastewater and stormwater in a hybrid constructed wetland system in southern Spain and its further reuse. Ecological Engineering 50:13-20.

Baptista J.D.C., Donnelly T., Rayne D., Davenport R.J. (2003) Microbial mechanisms of carbon removal in subsurface flow wetlands. Water Science and Technology 48(5):127-134.

Behrends L.L., Houke L., Bailey E., Jansen P., Brown D. (2001) Reciprocating constructed wetlands for treating industrial, municipal, and agricultural wastewater. Water Science and Technology 44(11-12):399-405.

Boog J. (2013) MS Thesis. Effect of the aeration scheme on the treatment performance of intensified treatment wetland systems. TU Bergakademie Freiberg, Freiberg, Germany.

Boog J., Nivala J., Aubron T., Wallace S., van Afferden M., Müller R.A. (2014) Hydraulic characterization and optimization of total nitrogen removal in an aerated vertical subsurface flow treatment wetland. Bioresource Technology 162:166-174.

Boutin C., Prost-Boucle S. (2015) Vertical flow constructed wetlands subject to load variations: an improved design methodology connected to outlet quality objectives. Water Science and Technology 72(5):817-823.

Brisson J., Chazarenc F. (2009) Maximizing pollutant removal in constructed wetlands: Should we pay more attention to macrophyte species selection? Science of the Total Environment 407(13):3923-3930.

Brix H. (1990) Gas exchange through the soil-atmosphere interphase and through dead culms of *Phragmites australis* in a constructed reed bed receiving domestic sewage. Water Research 24(2):259-266.

Brix H. (1994). Constructed wetlands for municipal wastewater treatment in Europe. In: W. J. Mitsch (ed.): *Global Wetlands: Old World and New*, Elsevier, Amsterdam, Netherlands, pp. 325–333.

Brix H. (1997) Do macrophytes play a role in constructed treatment wetlands? Water Science and Technology 35(5):11-17.

Brix H. (2017) Sludge dewatering and mineralization in sludge treatment reed beds. Water 9(3):160-172.

Brix H., Arias C.A. (2005) Danish guidelines for small-scale constructed wetland system for onsite treatment of domestic sewage. Water Science and Technology 51(9):1-9.

Brix H. and Johansen N. H. (2004). Guidelines for the Establishment of Reed Bed Systems up to 30 PE (Retningslinier for etablering af beplantede filteranlæg op til 30 PE). Økologisk Byfornyelse og Spildevandsrensning No.52, Miljøstyrelsen. Miljøministeriet, Copenhagen, Denmark (in Danish).

Butterworth E., Richards A., Vale P., Mansi G., Ranieri E., Dotro G., Jefferson B. (2016) Ammonia removal at four full-scale artificially aerated horizontal flow constructed wetlands. Water 8(9):365-380.

Canga E., Dal Santo S., Pressl A., Borin M., Langergraber G. (2011) Comparison of nitrogen elimination rates of different constructed wetland designs. Water Science and Technology 64(5):1122-1129.

Chernicharo C.A.L. (2007) Volume 4: Anaerobic Reactors. London, UK: IWA Publishing.

Cooper P.F., Job J.D., Green B., Shutes R.B.E. (1996) Reed beds and constructed wetlands for wastewater treatment, 184 pp. plus data diskette. Swindon, UK.

Corbella C., Garcia J., Puigagut J. (2016a) Microbial fuel cells for clogging assessment in constructed wetlands. Science of the Total Environment 569-570:1060-1063.

Corbella C., Garfi M., Puigagut J. (2014) Vertical redox profiles in treatment wetlands as function of hydraulic regime and macrophytes presence: surveying the optimal scenario for microbial fuel cell implementation. Science of the Total Environment 470-471:754-758.

Corbella C., Garfi M., Puigagut J. (2016b) Long-term assessment of best cathode position to maximise microbial fuel cell performance in horizontal subsurface flow constructed wetlands. Science of the Total Environment 563-564:448-455.

Corbella C., Puigagut J. (2015) Microbial fuel cells to improve treatment efficiency in constructed wetlands. In: 6th International Symposium on Wetland Pollutant Dynamics and Control (WETPOL) and Annual Conference of the Constructed Wetland Association (CWA), 13-18 September 2015. Cranfield University: York, UK. pp. 242-243.

CWA Database (2011) Constructed Wetlands Interactive Database. Compiled by the Constructed Wetland Association (CWA). Version 12.02.

Dal Santo S., Canga E., Pressl A., Borin M., Langergraber G. (2010) Investigation of nitrogen removal in a two-stage subsurface vertical flow constructed wetland system using natural zeolite. In: Proceedings of the 12th IWA Specialist Group

Conference on Wetland Systems for Water Pollution Control, 4-8 October 2010. Masi F., Nivala J. eds. IRIDRA S.r.l. and IWA: San Servolo, Venice, Italy. pp. 263-270.

Davison L., Headley T.R., Pratt K. (2005) Aspects of design, structure and performance and operation of reed beds-eight years' experience in northeastern New South Wales, Australia. Water Science and Technology 51(10):129-138.

DeMaeseneer J.L. (1997) Constructed wetlands for sludge dewatering. Water Science and Technology 35(5):279-286.

Dittmer U., Meyer D., Langergraber G. (2005) Simulation of a subsurface vertical flow constructed wetland for CSO treatment. Water Science and Technology 51(9):225-232.

Doherty L., Zhao Y., Zhao X., Hu Y., Hao X., Xu L., Liu R. (2015) A review of a recently emerged technology: Constructed wetland--Microbial fuel cells. Water Research 85:38-45.

Dotro G., Chazarenc F. (2014) Solids accumulation and clogging. Sustainable Sanitation Practice Journal 18:8-14.

Dotro G., Fort R., Barak J., Mones M., Vale P., Jefferson B. (2015) Long-term performance of constructed wetlands with chemical dosing for phosphorus removal. In: The Role of Natural and Constructed Wetlands in Nutrient Cycling and Retention on the Landscape. (ed) V.J., (ed. Springer Science + Business Media B.V.: Dordrecht, Netherlands. pp. 273-292.

DWA (2005) Empfehlungen für Planung, Konstruktion und Betrieb von Retentionsbodenfilteranlagen zur weitergehenden Regenwasserbehandlung im Misch- und Trennverfahren (*Recommendations for planning, construction and operation of retention soil filters for rainwater treatment in mixed and separated sewer systems)*. Deutsche Vereinigung für Wasserwirtschaft, Abwasser und Abfall e.V. (DWA): Hennef, Germany.

DWA (2017) Grundsätze für Bemessung, Bau und Betrieb von Kläranlagen mit bepflanzten und unbepflanzten Filtern zur Reinigung häuslichen und kommunalen Abwassers, in German. (*Principles of design, construction and operation of planted and unplanted filters for treatment of domestic wastewater*). Deutsche Vereinigung für Wasserwirtschaft, Abwasser und Abfall e.V. (DWA): Hennef, Germany.

Environmental Protection Agency Ireland (2010) Code of Practice: Wastewater treatment and disposal systems serving single households (PE less than or equal to 10).

European Commission (2008) Priority Substances and Certain Other Pollutants according to Annex II of Directive 2008/105/EC. Last accessed 13 July 2016: http://ec.europa.eu/environment/water/water-framework/priority_substances.htm.

Fan J., Liang S., Zhang B., Zhang J. (2013) Enhanced organics and nitrogen removal in batch-operated vertical flow constructed wetlands by combination of intermittent aeration and step feeding strategy. Environ Sci Pollut Res Int 20(4):2448-2455.

Fang Z., Song H.L., Cang N., Li X.N. (2013) Performance of microbial fuel cell coupled constructed wetland system for decolorization of azo dye and bioelectricity generation. Bioresource Technology 144:165-171.

Flores Rosell L. (2015) MS Thesis. Life Cycle Assessment of a constructed wetland system for wastewater treatment and reuse in Nagpur, India. Universitat Politècnica de Catalunya Barcelona, Department of Hydraulic, Maritime and Environmental Engineering. Barcelona, Spain.

Flyckt L. (2010) MS Thesis. Treatment results in Swedish Free water surface wetlands, operational experiences and their cost-efficiency for wastewater treatment, in Swedish. LITH-IFM-A-EX--10/2377-SE, IFM Biology, Linkoping University.

Foladori P., Ruaben J., Ortigara A.R. (2013) Recirculation or artificial aeration in vertical flow constructed wetlands: a comparative study for treating high load wastewater. Bioresource Technology 149:398-405.

Foladori P., Ruaben J., Ortigara A.R.C., Andreottola G. (2014) Batch feed and intermittent recirculation to increase removed loads in a vertical subsurface flow filter. Ecological Engineering 70:124-132.

Freeze R.A., Cherry J.A. (1979) Groundwater. Englewood Cliffs, New Jersey: Prentice-Hall.

Fuchs V.J., Mihelcic J.R., Gierke J.S. (2011) Life cycle assessment of vertical and horizontal flow constructed wetlands for wastewater treatment considering nitrogen and carbon greenhouse gas emissions. Water Research 45(5):2073-2081.

García J., Aguirre P., Mujeriego R., Huang Y., Oritz L., Bayona J.M. (2004) Initial contaminant removal performance factors in horizontal flow reed beds used for treating urban wastewater. Water Research 38:1669-1678.

García J., Corzo A. (2008) Purification with constructed wetlands: A practical guide to design, construction, and operating subsurface flow wetland systems. (*In Spanish: Depuración con Humedales Construidos: Guía Práctica de Diseño, Construcción y Explotación de Sistemas de Humedales de Flujo Subsuperficial*). Barcelona, Spain: Universitat Politècnica de Catalunya.

Garcia J., Rousseau D., Morato J., Lesage E.L.S., Matamoros V., Bayona J. (2010) Contaminant Removal Processes in Subsurface-Flow Constructed Wetlands: A Review. Critical Reviews in Environmental Science and Technology 40(7):561-661.

Green M., Friedler E., Ruskol Y., Safrai I. (1997) Investigation of alternative method for nitrification in constructed wetlands. Water Science and Technology 35(5):63-70.

Green M.B., Martin J.R., Griffin P. (1999) Treatment of combined sewer overflows at small wastewater treatment works by constructed reed beds. Water Science and Technology 40(3):357-364.

Gregersen P., Gabriel S., Brix H. and Faldager I. (2003) Guidelines for Willow Systems up to 30 PE. Økologisk Byfornyelse og Spildevandsrensning No.25, Miljøstyrelsen. Miljøministeriet, Copenhagen, Denmark (in Danish).

Griffin P., Wilson L., Cooper D. (2008) Changes in the use, operation and design of sub-surface flow constructed wetlands in a major UK water utility. In: Proceedings of the 11th International Conference on Wetland Systems for Water Pollution Control, 1-7 November 2008. Billore S.K., Dass P., Vymazal J. eds. Vikram University and IWA: Indore, India. pp. 419-426.

Haberl R., Grego S., Langergraber G., Kadlec R.H., Cicalini A.-R., Dias S.M., Navais J.M., Subert S., Gerth A., Thomas H., Hebner A. (2003) Constructed wetlands for the treatment of organic pollutants. Journal of Soils and Sediments 3(2):109-124.

Headley T.R., Tanner C.C. (2012) Constructed wetlands with floating emergent macrophytes: an innovative stormwater treatment technology. Critical Reviews in Environmental Science and Technology 42:2261-2310.

Higgins J.P. (2003) The use of engineered wetlands to treat recalcitrant wastewaters. In: Constructed Wetlands for Wastewater Treatment in Cold Climates. Mander Ü., Jenssen P., eds). WIT Press: Southampton, United Kingdom. pp. 137-160.

Hoffmann H., Platzer C., Winker M., von Muench E. (2011) Technology review of constructed wetlands: Subsurface flow constructed weltands for greywater and domestic wastewater treatment. http://www.susana.org/en/resources/library/details/930, last accessed 25 May 2017. Deutsche Gesellschaft für Internationale Zusammenarbeit (GIZ): Eschborn, Germany.

ISO 14040 (2006): Environmental management – Life cycle assessment – Principles and framework. International Organization for Standardization, Technical Committee: ISO/TC 207/SC 5 Life cycle assessment, https://www.iso.org/obp/ui/#iso:std:iso:14040:ed-2:v1:en (accessed 13 August 2017).

Jenssen P.D., Krogstad T., Paruch A.M., Mæhlum T., Adam K., Arias C.A., Heistad A., Jonsson L., Hellström D., Brix H. (2010) Filter bed systems treating domestic wastewater in the Nordic countries – Performance and reuse of filter media. Ecological Engineering 36(12):1651-1659.

Kadlec R.H. (2000) The inadequacy of first-order treatment wetland models. Ecological Engineering 15:105-120.

Kadlec R.H., Knight R.L. (1996) Treatment Wetlands, First Edition. Boca Raton, Florida: CRC Press.

Kadlec R.H., Wallace S.D. (2009) Treatment Wetlands, Second Edition. Boca Raton, Florida: CRC Press.

Khurelbaatar G., Sullivan C., van Afferden M., Rahman K.Z., Fühner C., Gerel O., Londong J., Müller R.A. (2017) Application of primary treated wastewater to short rotation coppice of willow and poplar in Mongolia: Influence of plants on treatment performance. Ecological Engineering 98:82-90.

Knowles P., Dotro G., Nivala J., García J. (2011) Clogging in subsurface-flow treatment wetlands: Occurrence and contributing factors. Ecological Engineering 37(2):99-112.

Knowles P.R., Davies P.A. (2009) A method for the in-situ determination of the hydraulic conductivity of gravels as used in constructed wetlands for wastewater treatment. Desalination and Water Treatment 1(5):257-266.

Knowles P.R., Griffin P., Davies P.A. (2010) Complementary methods to investigate the development of clogging within a horizontal sub-surface flow tertiary treatment wetland. Water Research 44(1):320-330.

Kuschk P., Wießner A., Kappelmeyer U., Weißrodt E., Kästner M., Stottmeister U. (2003) Annual cycle of nitrogen removal by a pilot-scale subsurface horizontal flow in a constructed wetland under moderate climate. Water Research 37:4236-4242.

Laber J., Perfler R., Haberl R. (1997) Two strategies for advanced nitrogen removal in vertical flow constructed wetlands. Water Science and Technology 35(5):71-77.

Langergraber G. (2017) Applying process-based models for subsurface flow treatment wetlands: Recent developments and challenges. Water 9(5):1-18.

Langergraber G., Haberl R. (2001) Constructed wetlands for water treatment. Minerva Biotechnologica 13(2):123-134.

Langergraber G., Leroch K., Pressl A., Rohrhofer R., Haberl L. (2009a) Experiences with a top layer of gravel to enhance the performance of vertical flow constructed wetlands at cold temperatures. Water Science and Technology 59(6):1111-1116.

Langergraber G., Prandtstetten C., Pressl A., Haberl R., Rohrhofer R. (2007) Removal efficiency of subsurface vertical flow constructed wetlands for different organic loads. Water Science and Technology 56(3):75-84.

Langergraber G., Pressl A., Leroch K., Rohrhofer R., Haberl R. (2011) Long-term behaviour of a two-stage CW system regarding nitrogen removal. Water Science and Technology 64(5):1137-1141.

Langergraber G., Rousseau D.P.L., García J., Mena J. (2009b) CWM1: a general model to describe biokinetic processes in subsurface flow constructed wetlands. Water Science and Technology 59(9):1687-1697.

Langergraber G., Simunek J. (2005) Modeling variably saturated water flow and multi-component reactive transport in constructed wetlands. Vadose Zone Journal 4:924-938.

Langergraber G., Šimůnek J. (2012) Reactive transport modelling of subsurface flow constructed wetlands using the HYDRUS wetland module. Vadose Zone Journal 11(2):14 pages.

Langergraber G., Sleytr K., Haberl R., Lindner G. and Zibuschka F. (2004). Experiences on using mineral wool as a drainage layer for vertical flow constructed wetlands. In: Proceedings of the 9th International Conference on Wetland Systems for Water Pollution Control, 26–30 September 2004, Lienard, A., Burnett, H. (eds), Avignon, France, pp. Paper No. PC3VF02 (CD-ROM).

Lauschmann R.E., Lechner M., Ertl T., Langergraber G. (2013) Experiences with pre-precipitation of phosphorus in a vertical flow constructed wetland in Austria. Water Science and Technology 67:2337-2341.

Lefebvre O., Uzabiaga A., Chang I.S., Kim B.-H., Ng H.Y. (2011) Microbial fuel cells for energy self-sufficient domestic wastewater treatment—a review and discussion from energetic consideration. Applied Microbiology and Biotechnology 89(2):259-270.

Li H.Z., Wang S., Ye J.F., Xu Z.X., Jin W. (2011) A practical method for the restoration of clogged rural vertical subsurface flow constructed wetlands for domestic wastewater treatment using earthworm. Water Science and Technology 63(2):283-290.

Li W.W., Yu H.Q. (2015) Stimulating sediment bioremediation with benthic microbial fuel cells. Biotechnology Advances 33(1):1-12.

Li Y., Zhu G., Ng W.J., Tan S.K. (2014) A review on removing pharmaceutical contaminants from wastewater by constructed wetlands: design, performance and mechanism. Science of the Total Environment 468-469:908-932.

Liénard A., Guellaf H., Boutin C. (2001) Choice of sand for filters used for secondary treatment of wastewater. Water Science and Technology 44(2-3):189-196.

Liquette C., Udias A., Conte G., Grizzetti B., Masi F. (2016) Integrated valuation of a nature-based solution for water pollution control. Highlighting hidden benefits. Ecosystem Services 22B(392-401).

Liu S., Song H., Li X., Yang F. (2013) Power generation enhancement by utilizing plant photosynthate in microbial fuel cell coupled constructed wetland system. International Journal of Photoenergy 2013:1-10.

Logan B.E., Hamelers B., Rozendal R., Schröder U., Keller J., Freguia S., Aelterman P., Verstraete W., Rabaey K. (2006) Microbial Fuel Cells: Methodology and Technology. Environmental Science and Technology 40(17):5181-5192.

Luo Y., Guo W., Ngo H.H., Nghiem L.D., Hai F.I., Zhang J., Liang S., Wang X.C. (2014) A review on the occurrence of micropollutants in the aquatic environment and their fate and removal during wastewater treatment. Science of the Total Environment 473-474:619-641.

Iwema A., Raby D., Lesarve J., Boutin C., Dodane P.-H., Liénard A., Molle P., Beck C., Sadowski G.A., Merlin G., Dap S., Ohresser C., Poulet J.-B., Reeb G., Werckmann M., and Esser D. (2005, eds): *Épuration des eaux usées domestiques par filtres plantes de macrophytes: Recommandations techniques pour la conception et la réalisation* (Treatment of domestic wastewater with planted filters: technical recommendations for design and realization). Groupe Français "Macrophytes et Traitement des Eaux" and Agence de l'Eau Rhône Méditerranée et Corse, et Rhin Meuse, France (In French). Available online: https://epnac.irstea.fr/wp-content/uploads/2012/08/Guide-Macrophytes.pdf.

Maltais-Landry G., Maranger R., Brisson J. (2009) Effect of artificial aeration and macrophyte species on nitrogen cycling and gas flux in constructed wetlands. Ecological Engineering 35(2):221-229.

Masi F., Bresciani R., Martinuzzi N. and Rizzo A.C.L. (2017a) Large scale application of French Reed Beds: municipal wastewater treatment for a 20 000 inhabitant's town in Moldova. Water Science and Technology, 76(1):134-146.

Masi F., Rizzo A.C.L., Bresciani R., Conte G. (2017b) Constructed wetlands for combined sewer overflow treatment: Ecosystem services at Gorla Maggiore, Italy. Ecological Engineering 98:427-438.

Matos M.P., Barreto A.B., Vasconcellos G.R., Matos A.T., Simões G.F., von Sperling M. (2017) Difficulties and modifications in the use of available methods for hydraulic conductivity measurements in highly clogged horizontal subsurface flow constructed wetlands. Water Science and Technology. doi: 10.2166/wst.2017.340.

McCuen R.H. (2016) Hydrologic Analysis and Design, Fourth Edition. Englewood Cliffs, New Jersey: Prentice Hall.

Mercoiret L. (2010). Qualité des eaux usées domestiques produites par les petites collectivités – Application aux agglomérations d'assainissement inférieures à 2 000 Equivalent Habitants. ONEMA and Cemagref. Lyon, France.

Meyer D., Chazarenc F., Claveau-Mallet D., Dittmer U., Forquet N., Molle P., Morvannou A., Pálfy T., Petitjean A., Rizzo A., Samsó Campà R., Scholz M., Soric A., Langergraber G. (2015) Modelling constructed wetlands: Scopes and aims – a comparative review. Ecological Engineering 80:205-213.

Meyer D., Dittmer U. (2015) RSF_Sim – A simulation tool to support the design of constructed wetlands for combined sewer overflow treatment. Ecological Engineering 80:198-204.

Meyer D., Molle P., Esser D., Troesch S., Masi F., Dittmer U. (2013) Constructed Wetlands for Combined Sewer Overflow Treatment – Comparison of German, French and Italian Approaches. Water 5:1-12.

Millot Y., Troesch S., Esser D., Molle P., Morvannou A., Gourdon R., Rousseau D. (2016) Design improvements for high ammonium removal by one-stage vertical flow constructed wetlands. Ecological Engineering 97:516-523.

Mitterer-Reichmann G.M. (2012) Treatment wetlands in Austria: Practical experiences in planning, construction and maintenance. Sustainable Sanitation Practice Journal 12:4-8.

Molle P., Liénard A., Boutin C., Merlin G., Iwema A. (2005) How to treat raw sewage with constructed wetlands: an overview of the French systems. Water Science and Technology 51(9):11-21.

Molle P., Liénard A., Grasmick A., Iwema A. (2006) Effect of reeds and feeding operations on hydraulic behaviour of vertical flow constructed wetlands under hydraulic overloads. Water Research 40(3):606-612.

Molle P., Lombard Latune R., Riegel C., Lacombe G., Esser D., Mangeot L. (2015) French vertical flow constructed wetland design: Adaptations for tropical climates. Water Science and Technology 71(10):1516-1523.

Morvannou A., Choubert J.-M., Vanclooster M., Molle P. (2014) Modeling nitrogen removal in a vertical flow constructed wetland treating directly domestic wastewater. Ecological Engineering 70:379-386.

Murphy C., Cooper D. (2011) An investigation into contaminant removal in an aerated saturated vertical flow constructed wetland treating septic tank effluent. In: Joint Meeting of Society of Wetland Scientists, WETPOL, and Wetlands Biogeochemistry, 3-8 July 2011. Vymazal J. (ed. Czech University of Life Sciences: Prague, Czech Republic. pp. 224.

Murphy C., Cooper D., Williams E. (2009) Reed bed refurbishment: A sustainable approach. 3rd International Symposium on Wetland Pollutant Dynamics and Control (WETPOL), 20-24 September 2009:193-194.

Murphy C., Wallace S., Cooper D. (2012) Treatment performance of two aerated saturated vertical flow constructed wetlands treating settled sewage. In: Proceedings of the 13th IWA Specialist Group Conference on Wetland Systems for Water Pollution Control, 25-29 November 2012. Mathew K., Dallas S. eds. Murdoch University, AWA and IWA: Perth, Australia, pp. 60.

Murphy C., Rajabzadeh A.R., Weber K.P., Nivala J., Wallace S.D., Cooper D.J. (2016) Nitrification cessation and recovery in an aerated saturated vertical subsurface flow treatment wetland: Field studies and microscale biofilm modeling. Bioresour Technol 209:125-132.

Murphy C., Wallace S., Knight R., Cooper D. and Sellers T. (2014). Treatment performance of an aerated constructed wetland treating glycol from de-icing operations at a UK airport. *Ecological Engineering*, 80, 117-124

NICNAS (2016) Priority Existing Chemical (PEC) Assessments https://www.nicnas.gov.au/chemical-information/pec-assessments (accessed 13 July 2016). National Industrial Chemical Notification and Assessment Scheme (NICNAS): Australia.

Nielsen S. (2012) Sludge Treatment in Reed Beds Systems – Development, design, experiences. Sustainable Sanitation Practice Journal 12:33-39.

Nielsen S., Willoughby N. (2005) Sludge treatment and drying reed bed systems in Denmark. Water and Environment Journal 19(4):296-305.

Nielsen S.M. (2003) Sludge drying reed beds. Water Science and Technology 48(5):101-109.

Nivala J. (2012) PhD Thesis. Effect of design on treatment performance, plant nutrition and clogging in subsurface flow treatment wetlands. Aarhus University, Department of Bioscience. Aarhus, Denmark.

Nivala J., Hoos M., Cross C., Wallace S., Parkin G. (2007) Treatment of landfill leachate using an aerated, horizontal subsurface-flow constructed wetland. Science of the Total Environment 380(1-3):19-27.

Nivala J., Knowles P., Dotro G., Garcia J., Wallace S. (2012) Clogging in subsurface-flow treatment wetlands: measurement, modeling and management. Water Research 46(6):1625-1640.

Nivala J., Murphy C., Troesch S., Wallace S. and Esser D. (2014). Intensified and modified wetland designs. *Sustainable Sanitation Practice Journal*, 18, 15–20.

Nivala J., Rousseau D.P.L. (2009) Reversing clogging in subsurface-flow constructed wetlands by hydrogen peroxide treatment: Two case studies. Water Science and Technology 59(10):2037-2046.

Nivala J., Wallace S., Headley T., Kassa K., Brix H., van Afferden M., Müller R. (2013) Oxygen transfer and consumption in subsurface flow treatment wetlands. Ecological Engineering 61:544-554.

O'Hogain S., McCarton L., Reid A., Turner J. (2010) A review of zero discharge wastewater treatment systems using reed willow bed combinations in Ireland. In: Proceedings of the 12th IWA Specialist Group Conference on Wetland Systems for Water Pollution Control, 4-8 October 2010. Masi F., Nivala J. eds. IRIDRA S.r.l. and IWA: San Servolo, Venice, Italy. pp. 755-762.

ÖNORM (2009) ÖNORM B 2505: Bepflanzte Bodenfilter (Pflanzenkläranlagen) - Anwendung, Bemessung, Bau und Betrieb (Subsurface flow constructed wetlands - Application, dimensioning, installation, and operation) [*in German*]. Österreichisches Normungsinstitut: Vienna, Austria.

Ouellet-Plamondon C., Chazarenc F., Comeau Y., Brisson J. (2006) Artificial aeration to increase pollutant removal efficiency of constructed wetlands in cold climate. Ecological Engineering 27:258-264.

Paing J., Guilbert A., Gagnon V., Chazarenc F. (2015) Effect of climate, wastewater composition, loading rates, system age and design on performances of French vertical flow constructed wetlands: A survey based on 169 full scale systems. Ecological Engineering 80:46-52.

Platzer C. (1999) Design recommendations for subsurface flow constructed wetlands for nitrification and denitrification. Water Science and Technology 40(3):257-264.

Platzer C., Hoffmann H., Miglio R.M. (2016) Long term experiences with dimensioning and operation of vertical flow constructed wetlands in warm climate regions of South America. In. International Water Association (IWA): Gdansk, Poland. pp. 868-887.

Prigent S., Paing J., Andres Y., Chazarenc F. (2013) Effects of a saturated layer and recirculation on nitrogen treatment performances of a single stage vertical flow constructed wetland (VFCW). Water Science and Technology 68(7):1461-1467.

Prost-Boucle S., Garcia O., Molle P. (2015) French vertical-flow constructed wetlands in mountain areas: How do cold temperatures impact performances? Water Science and Technology 71(8):1219-1228.

Prost-Boucle S., Molle P. (2012) Recirculation on a single stage of vertical flow constructed wetland: Treatment limites and operation modes. Ecological Engineering 43:81-84.

Rabaey K., Verstraete W. (2005) Microbial fuel cells: novel biotechnology for energy generation. Trends in Biotechnology 23(6):291-298.

Rani H.C.R., Din M.F.M., Yusof B.M., Chelliapan S. (2011) Overview of subsurface constructed wetlands in tropical climates. Universal Journal of Environmental Research and Technology 1(2):103-114.

Rousseau D.P.L., Vanrolleghem P.A., De Pauw N. (2004) Model-based design of horizontal subsurface flow constructed treatment wetlands: A review. Water Research 38:1484-1493.

Samsó R., Garcia J. (2013) BIO_PORE, a mathematical model to simulate biofilm growth and water quality improvement in porous media: Application and calibration for constructed wetlands. Ecological Engineering 54:116-127.

Samsó R., Garcia J., Molle P., Forquet N. (2016) Modelling bioclogging in variably saturated porous media and the interaction between surface/subsurface flows: Application to Constructed Wetlands. Journal of Environmental Management 165: 271-279.

Sherard J., Dunningan L., Talbot J. (1984) Basic properties of sand and gravel filters. Journal of Geotechnical Engineering Division, ASCE 110(6):684-700.

Stefanakis A., Akratos C.S., Tsihrintzis V.A. (2014) Vertical Flow Constructed Wetlands: Eco-engineering Systems for Wastewater and Sludge Treatment. Waltham, MA, USA: Elsevier.

Stefanakis A., Tsihrintzis V. (2012a) Effect of various design and operation parameters on performance of pilot-scale sludge drying reed beds. Ecological Engineering 38:65-78.

Stefanakis A.I., Tsihrintzis V.A. (2009) Effect of outlet water level raising and effluent recirculation on removal efficiency of pilot-scale, horizontal subsurface flow constructed wetlands. Desalination 248(1-3):961-976.

Stefanakis A.I., Tsihrintzis V.A. (2012b) Use of zeolite and bauxite as filter media treating the effluent of Vertical Flow Constructed Wetlands. Microporous and Mesoporous Materials 155:106-116.

Stein O.R., Borden-Stewart D.J., Hook P.B. and Jones W.L. (2007a) Seasonal Influence on Sulfate Reduction and Metal Sequestration in Sub-surface Wetlands. Water Research. 41(15):3440-3448.

Stein O.R., Towler B.W., Hook P.B., Biederman J.A. (2007b) On fitting the k-C* first order model to batch loaded sub-surface treatment wetlands. Water Science and Technology 56(3):93-99.

Strande L., Ronteltap M., Brdjanovic D. (2014) Faecal Sludge Management (FSM) Book - Systems Approach for Implementation and Operation: IWA, UK.

Sun G., Gray K.R., Biddlestone A.J., Cooper D.J. (1999) Treatment of agricultural wastewater in a combined tidal flow: Downflow reed bed system. Water Science and Technology 40(3):139-146.

Tanner C.C., D'Eugenio J., McBride G.B., Sukias J.P.S., Thompson K. (1999) Effect of water level fluctuation on nitrogen removal from constructed wetland mesocosms. Ecological Engineering 12(1-2):67-92.

Tanner C.C., Kadlec R.H. (2003) Oxygen flux implications of observed nitrogen removal rates in subsurface-flow treatment wetlands. Water Science and Technology 48(5):191-198.

Tao W., Bays J.S., Meyer D., Smardon R.C., Levy Z.F. (2014) Constructed Wetlands for Treatment of Combined Sewer Overflow in the US: A Review of Design Challenges and Application Status. Water 6:3362-3385.

Taylor C.R., Hook P.B., Stein O.R., Zabinski C.A. (2011) Seasonal effects of 19 plant species on COD removal in subsurface treatment wetland microcosms. Ecological Engineering 37(5):703-710.

Teixeira E.C., Siqueira R.d.N. (2008) Performance assessment of hydraulic efficiency indexes. Journal of Environmental Engineering 134(10):851-859.

Torrens A., Molle P., Boutin C., Salgot M. (2009) Impact of design and operation variables on the performance of vertical-flow constructed wetlands and intermittent sand filters treating pond effluent. Water Research 43(7):1851-1858.

Troesch S., Esser D. (2012) Constructed wetlands for the treatment of raw wastewater: The French experience. Sustainable Sanitation Practice Journal 12:9-15.

U.S. EPA (2000) Guiding principles for constructed treatment wetlands: Providing water quality and wildlife habitat. U.S. EPA Office of Wetlands, Oceans, and Watersheds.

U.S. EPA (2016) Toxic and Priority Pollutants Under the Clean Water Act. in: Agency U.S.E.P., ed. https://wwwepagov/eg/toxic-and-priority-pollutants-under-clean-water-act (accessed 13 July 2016).

Uggetti E., Ferrer I., Llorens E., García J. (2010) Sludge treatment wetlands: A review on the state of the art. Bioresource Technology 101(9):2905-2912.

Uggetti E., Ferrer I., Molist J., García J. (2011) Technical, economic and environmental assessment of sludge treatment wetlands. Water Research 45(2):573-582.

van Oirschot D., Wallace S.D., van Deun R. (2015) Wastewater treatment in a compact intensified wetland system at the Badboot: a floating swimming pool in Belgium. Environmental Science and Pollution Research 22(17):12870-12878.

Verlicchi P., Zambello E. (2014) How efficient are constructed wetlands in removing pharmaceuticals from untreated and treated urban wastewaters? A review. Science of the Total Environment 470-471:1281-1306.

Verlicchi P., Zambello E. and Al Aukidy M. (2014). Removal of personal care products in constructed wetlands. In: Personal Care Products in the Aquatic Environment, M. S. Díaz-Cruz and D. Barceló (eds), Springer International Publishing Switzerland, pp. 319–354.

Vohla C., Kõiv M., Bavor H.J., Chazarenc F., Mander Ü. (2011) Filter materials for phosphorus removal from wastewater in treatment wetlands—A review. Ecological Engineering 37(1):70-89.

von Sperling M. (2007a) Volume 1: Wastewater Characteristics, Treatment, and Disposal. London, UK: IWA Publishing.

von Sperling M. (2007b) Volume 2: Basic Principles of Wastewater Treatment. London, UK: IWA Publishing.

von Sperling M. (2007c) Volume 3: Waste Stabilisation Ponds. London, UK: IWA Publishing.

von Sperling M. (2007d) Volume 5: Activated Sludge and Aerobic Biofilm Reactors. London, UK: IWA Publishing.

von Sperling M. and Gonçalves R. F. (2007). Sludge characteristics and production. In: Sludge Treatment and Disposal, V. A. Cleverson, M. von Sperling, and F. Fernandes (eds), Biological Wastewater Treatment Series Volume 6, IWA Publishing, London, UK, pp.4-30.

Vymazal J. (1996) The use of subsurface flow constructed wetlands for wastewater treatment in the Czech Republic. Ecological Engineering 7:1-14.

Vymazal J. (1998). Czech Republic. In: J. Vymazal, H. Brix, P. F. Cooper, M. B. Green, and R. Haberl (eds), *Constructed Wetlands for Wastewater Treatment in Europe*, Backhuys Publishers, Leiden, The Netherlands, pp. 95–121.

Vymazal J. (2007) Removal of nutrients in various types of constructed wetlands. Science of the Total Environment 380(1-3):48-65.

Vymazal J. (2009) Horizontal sub-surface flow constructed wetlands Ondřejov and Spálené Poříčí in the Czech Republic – 15 years of operation. Desalination 246(1-3):226-237.

Vymazal J. (2011) Long-term performance of constructed wetlands with horizontal sub-surface flow: Ten case studies from the Czech Republic. Ecological Engineering 37(1):54-63.

Vymazal J. (2013) Emergent plants used in free water surface constructed wetlands: A review. Ecological Engineering 61B:582-592.

Vymazal J., Kröpfelová L. (2008) Wastewater Treatment in Constructed Wetlands with Horizontal Sub-Surface Flow: Springer.

Wahl M.D., Brown L.C., Soboyejo A.O., Martin J., Dong B. (2010) Quantifying the hydraulic performance of treatment wetlands using the moment index. Ecological Engineering 36(12):1691-1699.

Wallace S.D. (2001) Patent: System for removing pollutants from water. United States: US 6,200,469 B1.

Wallace S.D. (2014). Reducing wetland area requirements by using intensification strategies. In: Proceedings of the 14th IWA Specialist Group Conference on Wetland Systems for Water Pollution Control, 4-8 October 2010. IWA: Shangai, China. pp. 54-68.

Wallace S.D., Kadlec R.H. (2005) BTEX degradation in a cold-climate wetland system. Water Science and Technology 51(9):165-171.

Wallace S.D., Knight R.L. (2006) Small-scale constructed wetland treatment systems: Feasibility, design criteria, and O&M requirements. Water Environment Research Foundation (WERF): Alexandria, Virginia.

Wallace S.D., Liner M.O. (2011) Design and performance of the wetland treatment system at Buffalo Niagara International Airport. International Water Association's Specialist Group on Use of Macrophytes in Water Pollution Control No. 38:36-42.

Wallace S.D., Nivala J. (2005) Thermal response of a horizontal subsurface flow wetland in a cold temperate climate. International Water Association's Specialist Group on Use of Macrophytes in Water Pollution Control No. 29(February 2005):23-30.

WCED (1987). Our Common Future. Report of the World Commission on Environment and Development (Brundtland Report), Oxford University Press, Oxford, UK.

Wießner A., Kappelmeyer U., Kuschk P., Kästner M. (2005) Influence of redox condition on dynamics on the removal efficiency of a laboratory-scale constructed wetland. Water Research 39(1):248-256.

Wu S., Zhang D., Austin D., Dong R., Pang C. (2011) Evaluation of a lab-scale tidal flow constructed wetland performance: Oxygen transfer capacity, organic matter and ammonium removal. Ecological Engineering 37(11):1789-1795.

Yadav A., Dash P., Mohanty A.K., Abbassi R., Mishra B. (2012) Performance assessment of innovative constructed wetland-microbial fuel cell for electricity production and dye removal. Ecological Engineering 47(2012):126-131.

Zhao Y., Collum S., Phelan M., Goodbody T., Doherty L., Hu Y. (2013) Preliminary investigation of constructed wetland incorporating microbial fuel cell: Batch and continuous flow trials. Chemical Engineering Journal 229:364-370.

IWA Publishing's authorised EU representative for General Product Safety Regulations is Diane D'Arras, 15 rue Duret, 75116 Paris, France, e-mail: safety@iwap.co.uk.

Printed and bound by CPI Group (UK) Ltd, Croydon, CR0 4YY
12/05/2026
02108880-0008